Conveyor Belt Furnace Thermal Processing

Jinlong Xu • Joyce Zhang • Ken Kuang

Conveyor Belt Furnace Thermal Processing

Springer

Jinlong Xu
Hengli Eletek Co., Ltd
Hefei, Anhui, China

Joyce Zhang
Torrey Hills Technologies, LLC
San Diego, CA, USA

Ken Kuang
Torrey Hills Technologies, LLC
San Diego, CA, USA

ISBN 978-3-319-69729-1 ISBN 978-3-319-69730-7 (eBook)
https://doi.org/10.1007/978-3-319-69730-7

Library of Congress Control Number: 2017955960

Printed on acid-free paper

This Springer imprint is published by Springer Nature
The registered company is Springer International Publishing AG
The registered company address is: Gewerbestrasse 11, 6330 Cham, Switzerland

From Jinlong Xu
To the distinguished leaders at the 43rd Research Institute of CETC and the dedicated colleagues at Hengli Eletek Co., Ltd.

From Joyce Zhang
To Feng and Emma, for filling my life with love and joy.

From Ken Kuang
To my family, Zheng, Simon, and Andrew, you are my rock and inspiration!

Foreword

From the ancient Stone Age, to the Bronze Age, then to the Iron Age, the importance of thermal processing or heat treatment was gradually recognized by human beings. Very often the desired physical changes or chemical reactions required elevated temperatures. Thermal processing equipment that was invented to meet these needs gradually evolved from the ancient firewood fire, to kilns, ovens, and furnaces, to the present tokamak (a toroidal apparatus for producing controlled fusion reactions in hot plasma). The application of thermal processing started from the simplest form of clay sintering, to metallurgy, to the treating of modern advanced nano-materials. Thermal processing is one of the main means for human beings to transform natural materials and synthesize new materials.

The early development of the thermal processing technology was based on fire. With the inventions of electrical heating and advanced control, more and more modern thermal processing is done using electrically heated furnaces.

Due to the multitude of processing needs, most of the modern furnaces are customized. Among them, the industrial electric furnaces use electricity to heat up the material to a preset temperature to achieve desired physical and chemical changes. The operating principle of electrical heating is very simple, which is to convert electricity into heat energy through resistance, induction, radiation, or other heat transfer methods. A typical industrial electric furnace contains a heating system, a temperature control system, and in many cases an atmosphere system, when thermal processing requires controlled atmospheres such as air, nitrogen, hydrogen, vacuum, etc. A belt furnace also has some auxiliary systems, including transmission (loading) system and the encompassing furnace body frames.

There are mainly two types of industrial electric furnaces, namely, continuous and periodic. The continuous type has multiple temperature-controlled zones. It is able to continuously move the material by the use of conveyor belt, roller hearth, pushed slabs, or kiln car, driving the product through a required temperature profile. In comparison, the periodic type does not move the material while the surrounding temperature goes through a preset profile. Box oven, bell-type furnace, pit furnace, and shuttle kiln are good examples of periodic-type furnaces.

Currently, the photovoltaic, electronic components, metal heat treatment, powder metallurgy, and other related industries widely use continuous conveyor belt furnaces. Metal braided mesh belts are commonly used as a carrier in belt furnaces. After more than half a century of development, the first generation of mesh belt furnace with oxidative atmosphere heating has gradually developed to the second generation of protective, less or non-oxidative atmosphere heating, to the third generation of controlled atmosphere heating, and to the fourth generation of intelligent computer-controlled heating.

At the dawn of another wave of industrial revolution, intelligent, eco-friendly designs will be the development trend of all manufacturing equipment. Belt furnace is no exception.

The future of belt furnace design can be characterized by enhanced intelligence, flexible automation, high energy efficiency, cost reduction, and eco-friendliness. The future furnace will be an intelligent part of an automatic production line. The operator only needs to perform basic functions like inputting job requirement and uploading drawings into the computer, leaving the equipment system to be responsible for fabricating high-quality products on its own. The goal is to keep the design in line with RMS (reliability, maintainability, supportability) engineering philosophy while at the same time reducing unnecessary design cost and operation cost. Through the optimization of heating power and furnace atmosphere, assisted by smart PID precision algorithm control, facilitated by stable and reliable belt transmission, performance indicators such as furnace temperature, atmosphere uniformity, and system power will be automatically adjusted to the optimum level. The built-in self-cleaning system will treat exhaust gas to meet emission regulations, making the furnace green as it can be.

This book mainly introduces the conveyor belt furnace and its applications. We intend to share our experiences with readers about the performance, different processes, and the reliability of such equipment, including many detailed explanations in real-life settings. As we are keenly aware that no book can be perfect, what is presented to the readers is based on a collection of the authors' current understanding. Considering new technological development and process changes, equipment performance will continue to enhance, while furnace applications will further expand. I hope the authors will keep tracking the progress of thermal processing technology and provide readers with the latest updates in a future edition.

East China Research Institute of Microelectronics
Hefei, Anhui, China

Langping Li

Preface

Ever since the early 1990s, I started using different types of furnaces in a diverse range of industries, from passive component multilayer ceramic capacitor (MLCC), to ceramic packaging, to brazing, to solar cell manufacturing. Thermal processing with furnaces has played an important role in all of them. In 2004, as Torrey Hills Technologies, LLC, was still in its start-up phase, a major belt furnace manufacturer, Hengli Eletek, asked us to help market its furnaces internationally. I happily accepted and took up the challenge.

As we found out from the very beginning of this venture, the sales and marketing for high-temperature belt furnaces are not your typical sales pitch-based marketing. Our practices are technology oriented and they are focused on applications more than anything else.

As we started accessing clients in both familiar and unexpected fields of application, our team's knowledge accumulated over time, making us true experts in belt furnaces regardless of how and where the furnaces were used. This way we know we are doing the best possible job for our prospects and customers.

Over the last 13 years, we have installed belt furnaces in Australia, Austria, Brazil, Canada, People's Republic of China, Costa Rica, Hong Kong (China), Hungary, India, Indonesia, Israel, Italy, Malaysia, Mexico, the Netherlands, Russia, Singapore, Slovenia, South Korea, Spain, Sri Lanka, Switzerland, Slovenia, South Korea, Spain, Sri Lanka, Switzerland, Taiwan (Republic of China), Turkey, the UK, the USA, and Vietnam.

We are grateful our team has learned so much along the way.

As we were learning the different aspects of conveyor belt furnaces, we collected our understandings in many short articles (aka application notes) and put them on our website, with the hope that belt furnace designers, practitioners, and operators would find them useful. As expected, many peers expressed their appreciation for these application notes, and these articles were read and downloaded often.

This is my fifth book with Springer. The first two books, *Fuel Cell Electronics Packaging (2007)* and *RF and Microwave Microelectronics Packaging (2009)*, each have more than 15,000 downloads on Springer Link. When Marta Moldvai, editor at Springer Science+Business Media, asked me if we had a good idea for a new book, I shared our story and showed her the application notes. She was very happy to see them and encouraged us to add some more material and put them into a book so that more people can benefit. Thank you, Marta; without your encouragement, we wouldn't have come this far.

We are also very grateful to Mr. Dhanuj Nair (Springer India) and his team for the production assistance.

While putting together this book, we needed more chapters and invited a few friends to contribute some chapters. Their names were acknowledged in those chapters!

When the application notes were being written, we received generous assistance from many colleagues, like Nick Zhou, Ganesh Hariharan, and Michael Shaw, to name a few—thank you!

Torrey Hills Technologies, LLC
San Diego, California, USA

Ken Kuang

Contents

Introduction

Andrew Kuang

What Is Thermal Processing?

Thermal processing is defined as the combination of temperature and time required to achieve certain desired results. In ancient times, thermal processing via wildfire was first used to cook and preserve food. Soon it was used to harden/sinter clay and then in the metal ore smelting. This high-temperature treatment has been evolving ever since, and it is widely used in both food processing and many industrial high-temperature processing today.

This book focuses on the industrial high-temperature treatment.

In today's real-life applications, high-temperature treatment (aka heat treating or heat treatment) includes a group of industrial and metalworking processes used to alter the physical, and sometimes chemical, properties of a material.

Different Types of Thermal Processing

1. Batch Process

For small-scale lower-volume applications, batch furnaces are used for thermal processing. Batch furnaces can be used in small-scale operations, where a high production throughput is not needed. A batch furnace typically consists of an insulated chamber, a door, and a heating component that may be powered electrically or by using gas. Different types of batch furnaces include bell, box, pit, or car bottom.

2. Continuous Process

Many larger-scale applications requiring a high production rate may require the use of a continuous furnace. Some of the smaller-scale processing needing specific heating and cooling profile prefer a continuous furnace too. Continuous furnaces

J. Xu et al., *Conveyor Belt Furnace Thermal Processing*,
https://doi.org/10.1007/978-3-319-69730-7_1

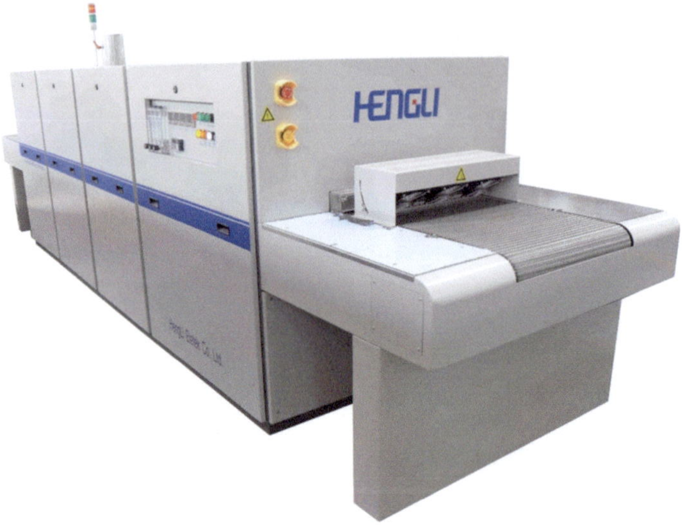

Fig. 1 A typical conveyor belt furnace (Photo courtesy of Hengli Eletek, Co. Ltd)

Fig. 2 A typical pusher kiln (Photo courtesy of Hengli Eletek, Co. Ltd)

constantly move materials through a heated chamber and comprise of a heating element, a conveyor system, and a heated compartment. Types of continuous furnaces include:

(i) Conveyor belt furnaces (Fig. 1)

(ii) Pusher kilns (Fig. 2)

Fig. 3 A typical roller hearth furnace (Photo courtesy of Hengli Eletek, Co. Ltd)

(iii) Roller hearth (Fig. 3)

(iv) Others

What Is a Conveyor Belt Furnace?

A conveyor belt furnace is a device used for rapid heating of substances, using a conveyor belt to move materials through a heated chamber.

Basic Structure of Conveyor Belt Furnace

Figure 4 shows a simplified structure of a conveyor belt furnace. The structure can be divided into a few key components:

(i) Loading and unloading tables
(ii) Heated chamber
(iii) Cooling chamber
(iv) Temperature control system
(v) Belt driving system (Table 1)

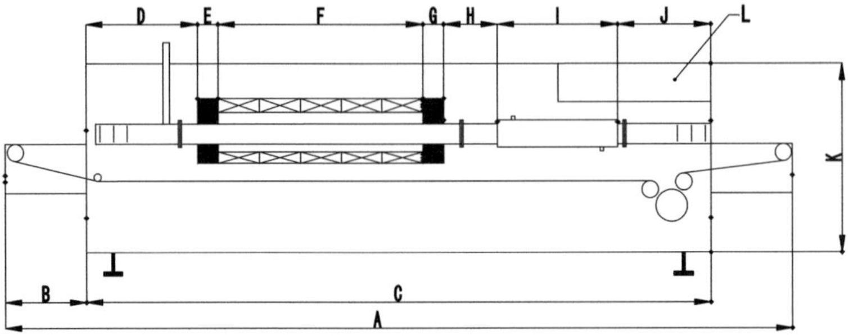

Fig. 4 A typical conveyor belt furnace layout (Courtesy of Hengli Eletek, Co. Ltd.)

Table 1 The detailed description of each section

No.	Name	Note
A	Total length	From 0.5 m like a table top bread belt toaster to over 20 m long
B	Load and unload table	The height is typically at 800 mm, for operators or robotic arms to load and unload parks after thermal processing
C	Furnace frame	This is main body for a furnace
D	Entrance vestibule zone	This section includes entrance curtain (multilayer stainless steel sheet or silicon rubber curtains). This is the important transition zone to ensure the inside furnace atmosphere is separated from outside atmosphere
E	Entrance insulation zone	Ceramic fiber insulation and transition to the heated section
F	Heated zone	The zones are heated in a controlled fashion. This section can have only one zone or multiple zones
G	Exit insulation zone	Ceramic fiber insulation and transition to the cooling section in a controlled way
H	Free cooling zone	Gas inlet and test welding reserve place
I	Water cooling zone	Some furnaces have this feature of allow increased cooling rate
J	Exit vestibule zone	This section includes entrance curtain (multilayer stainless steel sheet or silicon rubber curtains). This is the important transition zone to ensure the inside furnace atmosphere is separated from outside atmosphere
K	Furnace height	Exclude the mounting feet and exhaust stack height
L	Equipment operation panel	Typically the main control panel/machine and human interface are located here

As discussed above, very often a belt furnace's inside atmosphere is separated from outside atmosphere. This way is to ensure that the parts are processed in desired atmosphere (air, H_2/N_2, N_2, Ar, and controlled moisture content/dew point).

Below shows a simplified atmosphere control system design.

In this simplified design, there are eight atmosphere pipes. A, B, C, and D: the measuring range of five nitrogen flowmeters is 12–120 l/min, and one nitrogen

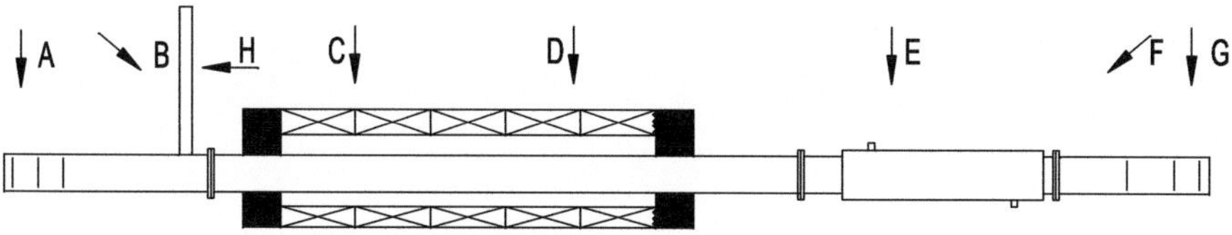

Fig. 5 Simplified gas layout drawing (Courtesy of Hengli Eletek, Co. Ltd)

Table 2 Simplified furnace atmosphere control system

Code	Name	Explanation
A	Entrance curtains	Import nitrogen curtain, prevent outside air from entering the chamber
B	Entrance gas inlet	Nitrogen purge products. Prevent outside air from entering the chamber
C	Heating gas inlet	Used for normal nitrogen supplement of heating zone
D	High-temperature gas inlet	Used for nitrogen protection and normal nitrogen supplement of firing zone
E	Cooling gas inlet	Used for normal nitrogen supplement and cooling parts of cooling zone
F	Exit gas inlet	Nitrogen purge products. Prevent outside air from entering the chamber
G	Exit curtains	Exit nitrogen curtain, prevent outside air from entering the furnace to ensure the product of the firing process requirements
H	Venturi exhaust gas inlet	The manual valve on Venturi exhaust tube can be used for adjusting gas flow

The actual gas flow of each pipe can be displayed by flowmeters

Venturi exhaust flowmeter has 5–50 L/min range. Please see Fig. 5 and Table 2 for detailed gas layout.

The temperature control system has three main components:

(a) Temperature measurement instrument: typically, thermocouples are used.
(b) Temperature control: use of intelligent temperature controller to control, with PID parameter auto-tuning, high-temperature upper limit alarm, thermocouple failure, and other protective alarm functions.
(c) Heating control system: for the heating control, grouped supply the electricity by three alternating current contacts to reduce the instantaneous current impact to the electric fence caused by the equipment.

The belt drive system design uses the variable frequency drive to control the speed by gear box. Typically, the speed is adjustable continually. There is a belt deviation guide device on the both sides at the entrance and exit.

Typical Applications

1. Water/Solvent Drying

Conveyor belt dryers are used very frequently in drying water as in the food processing industry or solvents as in screen printing.

2. Surface Mount

Reflow soldering of surface mount electronic components to printed circuit boards (PCBs) is typically done in a belt furnace too.

3. Die Attach

A die attachment is a process where a target slice of semiconductor is picked from the waffle and mounted on a substrate or a die bond table. Typically, die bonding technologies include eutectic bonding, epoxy bonding, and solder attach. Heating at specific temperatures in a belt furnace forms the bonding, cures the glue, and shapes the solder between the dies and the packages to improve mechanical, thermal, and electrical properties of the products.

4. Brazing

Another usage of a conveyor belt furnace is in brazing. When using a furnace to join using a brazing technique, filler is placed at the joint. Then, the component is fed into the furnace so the filler can create the joint. Using a belt furnace for brazing allows for the capability to join numerous parts at once, a high production rate, and the ability to closely regulate the temperature at which the joints will be brazed.

5. Direct Bond Copper (DBC)

The outstanding thermal conductivity makes DBC substrates prevalent in power modules. The basic structure of DBC consists of a ceramic substrate (Al_2O_3, BeO, or AlN), and a sheet of copper is bonded to the substrates by a high-temperature oxidation process on one or both sides. Heating the copper and substrate to a specifically controlled temperature in an atmosphere of nitrogen with about 30 ppm of oxygen forms a copper-oxygen eutectic between the copper and substrate. The top copper layer is then chemically etched to form an electrical circuit by printed circuit board technology. The bottom copper layer, on the other hand, is usually kept whole and soldered to the substrate to attach a heat spreader (Fig. 6).

6. MLCC, Capacitors and More Thick Film

A series of ceramic capacitors assembles together to make an MLCC. A thin ceramic foil is divided into small sheets with same size and serves as the electrodes. The number of layers and solidity of the sheets decides the capacitance value. The electrodes are gathered by the binder which is later burnt out. The burning process also determines the dielectric properties of MLLC. The sintering step is typically done in belt furnaces (Fig. 7).

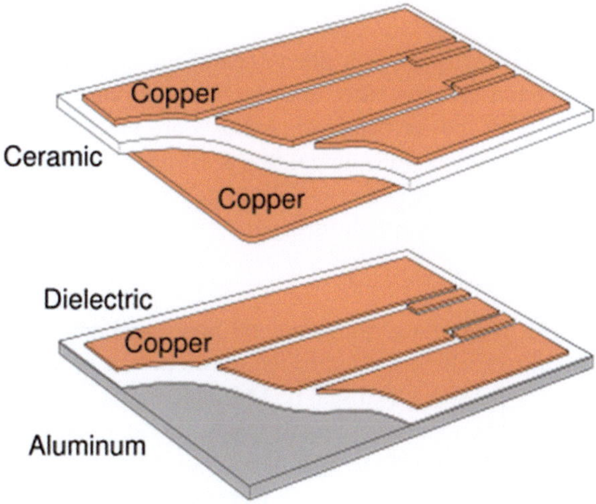

Fig. 6 Structure of a direct bonded copper substrate (top) and an insulated metal substrate (bottom)

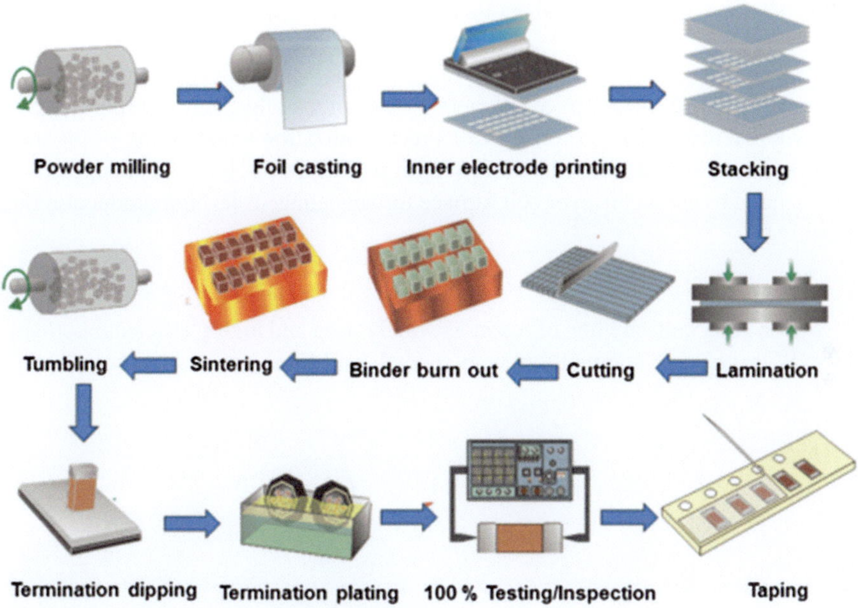

Fig. 7 Manufacturing process of MLLC

7. Automotive Thick Film Heaters

Thick film technology is used to create a number of different devices, especially parts used in the automotive industry. When creating parts, different substrates, particularly alumina, are machined using a laser. Ink is then transferred onto the substrate using a screen printing technique. The ink is then dried at a temperature between 50 and 300 °C, using a conveyor belt furnace.

8. Piezoelectric

Piezoelectricity is a phenomenon of mechanical energy and electrical energy converting to each other in dielectric material. Naturally, the positive and negative charges of the ferroelectric ceramics are not balanced. After burning process, the ferroelectric ceramics is polarized, forms permanent electrodes, and is transferred into piezoelectric ceramics. The metallization for piezoelectric ceramics is typically done in a conveyor belt furnace.

9. Glass-to-Metal Seal

The glass, under suitable conditions, will bond well to a wide variety of metals and alloys which has led to the development of many useful technologies. One of them is glass-to-metal seal (GTMS) technique. GTMS becomes a popular and low-cost method to hermetic sealing, which is an important part of design of many products today. The GTMS is a fusion bonding technique; glass is heated to a molten state and bonds to the metal. Bonding occurs through wetting and chemical reactions between the metal and the glass. In addition to providing seal, the glass acts as an insulator between one pin to another and housing.

10. Annealing

Conveyor belt furnaces have a variety of uses in metallurgy. Annealing is a process where a material is heated above its recrystallization temperature, so the metal's ductility will increase, while its strength will decrease. Large-scale operations will sometimes use a conveyor belt furnace for annealing, as its high production rate is optimal.

11. Silicon Solar Cell

Silicon solar cells are one of the most widely used and highly efficient photovoltaics. With growing competition from second- and third-generation photovoltaics

Fig. 8 Silicon solar cell

and decreasing module prices, there is an increasing demand for improving the efficiency and decreasing the cost of the silicon solar. For conventional silicon solar cells, the metallic stripes (electrodes) are screen printed and sintered in a belt furnace (Fig. 8).

12. Dye-Sensitized Solar Cell

Figs. 9 and 10 Dye-sensitized solar cell

Dye-sensitized solar cells are the most cost-effective, third-generation solar technology available today. DSSC technology, invented in 1991 by Michael Graetzel and Brian O'Regan, has recently attracted more interest because of its low material cost, ease of production, and high conversion efficiency compared to other thin-film solar cell technologies. In the manufacturing of DSSC, TiO_2 is screen printed onto the glass panel and sintered onto the glass in a belt furnace (Figs. 9 and 10).

13. Coin Annealing

Coins are widely used worldwide as a currency. They are typically produced in the following steps:

(a) Stamping—Round disks of metal are punched out of a long sheet. The disks, known as blanks, have no designs pressed onto them.
(b) Annealing—Coins are put through an annealing furnace to soften the metal, so designs can be more easily stamped on. Conveyor belt furnaces with H2/N2 atmosphere are often used in the annealing step, as their high production rate is best suited to producing coins.
(c) The annealed coins are then washed in large, rotating barrels with chemicals and then dried.
(d) Next, blanks are fed into an upsetting machine, which forms the edge which is found on nickels. The formed edge helps with stamping of designs.
(e) Then, coin blanks are fed into coin presses, where they are stamped with various designs and engravings.
(f) Lastly, the coins are inspected for defects and are then bagged and sent to the central banks (Fig. 11).

Fig. 11 Coin annealing products

Fig. 12 Cadmium telluride (CdTe) solar cells

14. Thin-Film CdTe Solar Cell Deposition Annealing

Cadmium telluride (CdTe) solar cells are particularly based on photovoltaic (PV) technology. By using thin film of cadmium telluride, the solar cells are able to generate electricity in a more efficient way than silicon solar cells do. In the manufacturing process, CdTe is deposited onto a large glass panel, and a conveyor belt furnace is used to anneal the deposited thin film. CdTe is typically a thin film deposited onto a big glass panel. After deposition, a stress relief process step is needed and is typically done in a belt furnace (Fig. 12).

Understanding the Influence of Belt Furnace and Firing Parameters on Efficiency of Silicon Solar Cells

Introduction

The metallization contact formation is an important step in determining the efficiency of the solar cell. The common metallization processes include photolithography, buried contact technology, and screen printing. The photolithography process involves selectively etching patterns of thin-film material on a silicon substrate. Photolithography is widely used in laboratories and produces highest-efficiency solar cells. However, this method is both time and resource intensive limiting it to laboratory research. The other more efficient metallization contact formation method is the buried contact method. In this method, grooves are scribed on the silicon wafer using mechanical or laser methods, and copper is plated in the holes. The contact metallization is achieved through electroless plating of Ni/Cu followed by sintering. The better aspect ratio obtained from the grooving helps in increasing the efficiency of solar cell. The third and most commonly used method in commercial production is the screen printing method. Screen printing is a cost-effective method that uses a screen and squeegee for printing the metallization paste onto the substrate which is then fired for contact formation. This work will focus on summarizing the ideal screen printing and firing process parameters and their impact on cell efficiency.

Screen Printing

The screen printing method consists of a thick-film metal paste that is composed of metal powder, glass frit, solvent, and nonvolatile polymers that are blended together in a three-roll mill. A squeegee applies a downward force on the paste moving it across the screen that has a deposition pattern in it. The action creates a reduction in viscosity of the paste making it penetrate into the screen holes. This way the metal paste is deposited in select patterns onto the substrate. The factors that affect the screen printing process include snap-off distance, squeegee pressure, and squeegee speed.

J. Xu et al., *Conveyor Belt Furnace Thermal Processing*,
https://doi.org/10.1007/978-3-319-69730-7_2

The snap-off distance is the distance between the screen and the wafer. During the printing process, when the paste is printed on the screen, a downward force is applied on the screen. The screen being elastic restores its shape back, and this upward movement aids in deposition of the paste. If the snap-off distance was too high, too much pressure will have to be applied to force the paste onto the wafer. If it is too low, the paste might not get released from the screen. The pressure applied also plays an important role in deposition of the paste. When too much of a pressure is applied, excessive paste from the screen could be deposited, and this could break the wafer. On the contrary, when too little pressure is applied, the paste might not get released from the screen. The speed of the squeegee movement also determines the print quality. When the speed is too high, the paste misses many holes leading to nonuniform deposition.

Firing

The firing process also referred to as sintering is one of the key steps during which the front metal contact is formed in a silicon solar cell. In this process, the thick-film paste is dried at about 150 °C to remove much of the solvents. The presence of solvents can cause excessive outgassing leading to cracks and voids. The dried substrates are then fired in a firing furnace. The firing process consists of four primary steps. The first step is the initial temperature ramp-up during which the paste solvents are volatilized. The second step is the burnout. The objective of the burnout phase is to remove all of the organic binder that was used in paste formation. The burnout phase is carried out at 300–400 °C. The third important step is the sintering or firing process which is done at 700–800 °C range. During this process, the Ag metal forms a bonding with the underlying silicon substrate to form metal contact. The final step is the wafer cooldown.

The mechanism behind contact formation in a fire-through contact is very complex and not fully understood [Ferro]. According to Mohhamed et al., the process starts with evaporating the solvent between 100 and 200 °C and then burning out the polymer between 200 and 400 °C. Later from 400 to 600 °C, the glass frit melts and the sintering of the Ag particles also takes place. Further from 600 to 800 °C, molten glass with some amount of dissolved Ag etches the silicon nitride antireflection coating and then reaches the Si surface. Here it reacts and etches a very thin layer of Si. Ag in the glass then precipitates onto the Si surface in the form of crystallites. An ideal temperature profile for the firing process is illustrated in Fig. 1.

The quality of the contact influences the shunt resistance, series resistance, and junction leakage current which have a significant impact on the efficiency of the solar cell. Hence it is very important to understand various aspects of the firing process to be able to achieve higher efficiencies. Some of them include peak temperature, dwell time, and the temperature at which Si-Ag alloy formation happens.

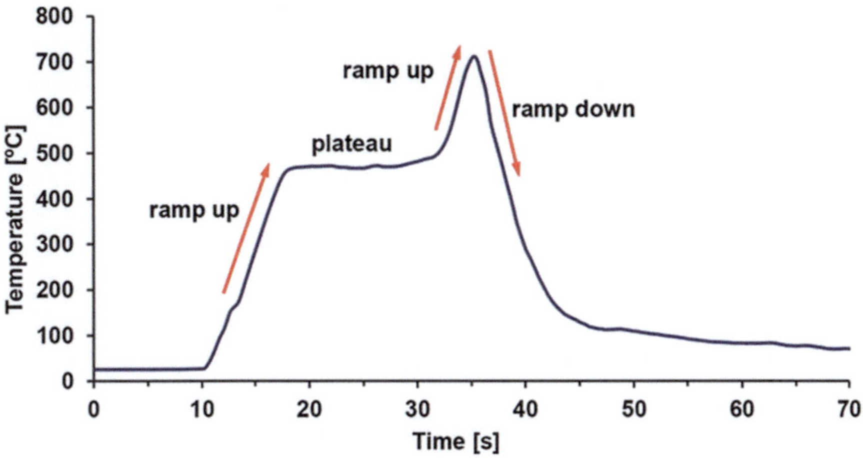

Fig. 1 Ideal temperature profile for contact firing [1]

Effect of Peak Temperature

The peak temperature of the sintering process is observed to have a significant impact in determining the efficiency of the silicon solar cell. Studies by Carroll et al. have reported a direct correlation between peak temperature and cell efficiency. From the results of this study, it can be observed that the ideal peak temperature range for metallization firing is 780–800 °C. Also, Cheng et al. have studied the effect of various peak temperatures on the efficiency of solar cell and confirmed the same. The study has observed that when silicon solar cell is underfired, one half of the cells display high contact resistance, while the other half displays low contact resistance. At the same time, when the solar cells are fired in the optimum range of 780–800 °C, the contact resistance has been reported to be uniform throughout with higher efficiency. However, when the solar cells were overfired, Cheng et al. observed overall decrease in efficiency due to shunting. Also investigation by Hilai et al. has demonstrated similar phenomena. At lower temperatures, it was observed that the distribution of Ag was irregular and the contact formed was very small with high specific contact resistance. However, at high temperatures, the excessive shunting resulted in decrease in efficiency. Thus from the results of the above studies, it can be observed that the peak firing temperature has an important role in determining the efficiency of the silicon solar cells.

Effect of Soak Time

Soak time refers to the time for which the substrates with metallization printed are held at peak temperature. The soak time is an important parameter in determining the cell efficiency as the diffusion of silicon and silver is a function of time and temperature. Research by Lee et al., Cooper et al., and Hilai et al. have demonstrated the effect of soak time on the efficiency of the solar cell. From the results of the published literature, it can be observed that there is a significant interaction between soak time and peak temperature. The highest efficiencies have always been reported for shorter soak time. It has been reported that increasing the soak time increased the contact resistance. Study of the interface and bulk metallurgies of the contact by Cooper et al. has demonstrated an increase in the thickness of interfacial glass layer with increasing soak time. The thicker contact layer results in increased contact resistance.

Belt Furnace Parameters

Apart from the firing conditions, studies in literature have reported a correlation between the furnace parameters and the efficiency of the solar cell. Edward Bruce has studied the influence of belt furnace parameters on the solar cell efficiency. It has been observed that the cleanliness of the furnace has a significant impact on the cell efficiency. The common source for furnace impurities includes carbon residue, impurities from air inlet, foreign materials picked up by the conveyor due to contact, etc. Solar cells fired immediately after cleaning the furnace have been reported to be more efficient than their counterparts that have been fired on furnace with impurities. Also, the thickness of the oxide layer has been reported to have a significant impact on the cell efficiency. The oxide layer thickness can be controlled by controlling the atmosphere inside the muffle. Hilai et al. have reported an increase in conductivity of silicon solar cell fired in a reducing atmosphere with small percentage of hydrogen. The substrate carrier boat has been reported to have very little impact in determining the cell efficiency. Results of Bruce et al. have shown no significant impact in efficiency for directly placing the silicon substrate on the nichrome belt and firing them on a quart carrier.

Furnace for Firing Operation

The HSH series furnace is a specially designed infrared furnace to cater to the needs of photovoltaic metallization firing requirements. The heating in this furnace is achieved with the help of shortwave infrared lamp heaters. The fast response of the IR lamps allows quick heating. The furnace is rated at 1000 °C and can operate very well in the 750–800 °C range required for sintering of front contact metallization. The belt width comes in various standard sizes including 250 mm, 300 mm, and 380 mm to match with the requirements of the wafer size. Cooling can be achieved through forced air as well as water per requirement. The presence of a muffle in the furnace helps in controlling the atmosphere inside the furnace as well as preventing external atmosphere from entering inside. In short, the muffle design helps maintain a cleaner furnace atmosphere. Also, this furnace is equipped with a steel brush as a standard feature. The steel brush helps in cleaning the conveyor belt. Ultrasonic belt cleaning is available as an extra option.

The furnace is controlled by microprocessor-based PID controller. Type K thermocouples are used for determining the zone temperatures. Controls are located on the right-hand side as viewed from the entrance of the furnace. The central processing unit (CPU) is mounted under the exit table. The furnace is controlled by a microprocessor-based controller system. The CPU is loaded with a Windows operating system that allows easy computing. The computer system comes with a preinstalled program for controlling the Confurnace parameters including belt speed and zone temperatures. In temperature profiles, one set can be stored and retrieved for future purposes. Thermocouple ports are located at the entrance table to connect the profiling thermocouple directly to the microprocessor. This feature allows monitoring and recording the actual temperature experienced by the part. Software is provided with the computer to capture, display, print out, and store furnace profile. The furnace is equipped with a redundant overheat safety protection system which incorporates an additional type "K" thermocouple in the center of each controlled zone and a multi-loop alarm.

Detailed specification for a HSH2503-0509 furnace is available in Fig. 2 and Table 1.

Table 1 Specifications of HSH2503-0509 firing furnace

Specification	HSH2503-0509
Rate temperature	900 °C max normal operating: 850 °C
Belt width	10″ (250 mm)
Effective above belt clearance	0.035″ (9 mm)
Control zones	5
Conveyor speed	28–140 IPM (700–3500 mm/min)
Muffle opening size	11 × 1.2″ (W × H) 280 × 30 mm (W × H)
Loading table	18 × 33″ (LXH) 450 × 850 mm (LXH)
Unloading table	22 × 33″ (LXH) 550 × 850 mm (LXH)
Belt	Balanced V weave, Cr20Ni80
Heating elements	Shortwave IR lamps
Insulation	High-quality ceramic fiber
Temperature controller	Intelligent PID Shimaden Controller
Alarm	Thermocouple, over temp, belt stop. Audio and visual alarm
Atmosphere	Four pipes of dry clean air or N_2. 2–6 m^3/h, 1.1–3.3 CFM
Cooling	Two-stage water cooling. 2–3 m^3/h, incoming water at 10 °C or colder
Across-belt temperature uniformity	±4 °C
Overall system width	40″ (1000 mm)
Overall system length	14′/168″/4290 mm
Overall system height	52″ (1316 mm)
Net weight	1760 lb. (800 kg)
Power	Three-phase, 480VAC, 60 Hz, 38 KVA max normal operating power draw is about 15 KVA

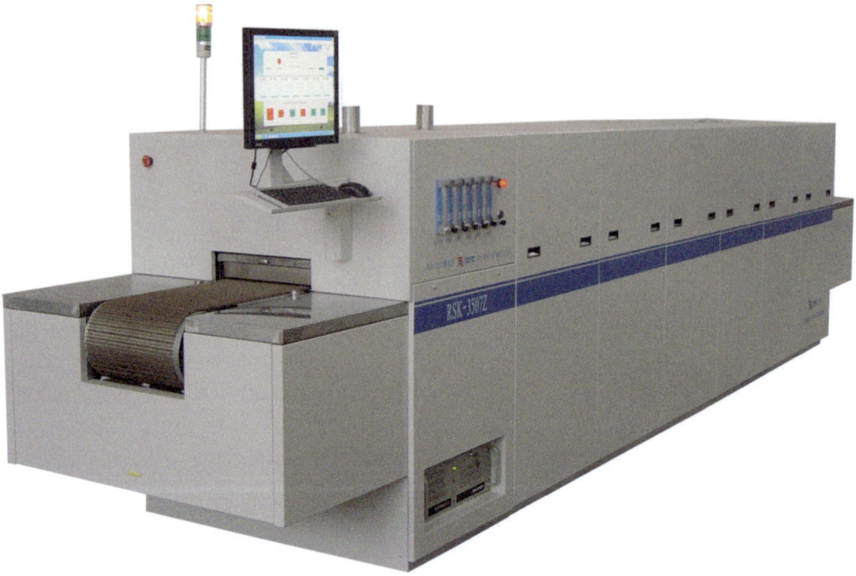

Fig. 2 HSH2503-0509 firing furnace

Conclusion

Firing process is a very important step in determining the efficiency of the silicon solar cell. The peak temperature of the firing, the soak time, and the furnace design itself has a significant influence on the efficiency of the silicon solar cell. The HSH series is an IR furnace that matches well with the requirements for this firing operation.

References

1. Sopori, B., & Mehta, V. (2007). Fundamental mechanism in the fire through contact metallization of silicon solar cells: A review. In *17th workshop on crystalline silicon solar cells and modules: Materials and processes*, Aug 508.
2. Cooper, I.,. B., & Ebong, A. (2010, May). Understanding and use of IR belt furnace for rapid thermal firing of screen-printed contacts to Si solar cells. *IEEE Electron Device Letters, 31*(5), 461–463.
3. Edwards, B. M. (2008). *Screen and stencil print technologies for industrial N-Type silicon solar cells*. PhD thesis, University of New South Wales.
4. Hilali, M. M., & Al-Jasim, M. M. (2005). Understanding the formation and temperature dependence of thick-film Ag contacts on high-sheet-resistance Si emitters for solar cells. *Journal of Electrochemical Society, 152*(10), 742–749.

5. Hoornstra, J., & Heurtault, B. (2009). Stencil print applications and progress for crystalline solar cells. In *24th European photovoltaic solar energy conference and exhibition*, 21–25 Sep, Hamburg.
6. Hilali, M. (2005). *Understanding and development of manufacturable screen printed contacts on high sheet resistance emitters for low cost silicon solar cells*. PhD dissertation, Georgia Institute of technology.
7. Mihailetchi, V. D., & Jourdan, J. *Screen printed N Type silicon solar cells for industrial applications*. In 25th European Photovoltaic Solar Energy Conference and Exhibition/5th World Conference on Photovoltaic Energy Conversion, 6-10 September (2010), Valencia, Spain.
8. Ebong, A., Brody, J., Rohatgi, A., & Williams, T. (1999). Optimization of front metal contact firing scheme to achieve high fill factors on screen printed silicon solar cells. In *PVSEC 1999 Japan*.

Understanding the Influence of Belt Furnace and Firing Parameters on Efficiency of Thin-Film CIGS Solar Cells

Introduction

The ever-increasing demand and growing cost of silicon has made thin-film photovoltaics a competing substitute to silicon solar. Thin-film photovoltaic (PV) modules are seen as a realistic alternative for a cost-effective generation of electricity from sunlight. They are often referred to as the second generation of photovoltaic technology. Some promising materials for thin-film solar cells include amorphous silicon, cadmium telluride, and CuInSe2 and its alloys. While all of these materials are low bandgap, they are all polycrystalline as well leading to a loss in efficiency due to grain boundary recombination. Amid the polycrystalline thin-film solar, CIGS solar cells have been documented in various pieces of literature as being unrivaled in efficiency.

A typical fabrication of a CIGS solar cell encompasses the sputtered deposition of molybdenum back contact material on a substrate. The substrate material can be a rigid soda lime glass or a flexible polyimide. The sputtering of the back contact is followed by the deposition of the CIGS absorber layer, CDS buffer layer, and the final ZnO contact layer. Molybdenum is primarily used as a back contact due to its ability to form an ohmic contact, as well as inert behavior to corrosive gases. Cadmium sulfide is used with CIGS material to form an n-type semiconductor material along with a p-type CIGS absorber material. Some of the widely used methods for depositing CDS include chemical bath deposition (CBD), sputtering, and closed-space sublimation (CSS). ZnO is used as the front contact material due to its superior electrical and optical properties. Figure 1 illustrates a material stack up of the CIGS solar cell.

J. Xu et al., *Conveyor Belt Furnace Thermal Processing*,
https://doi.org/10.1007/978-3-319-69730-7_3

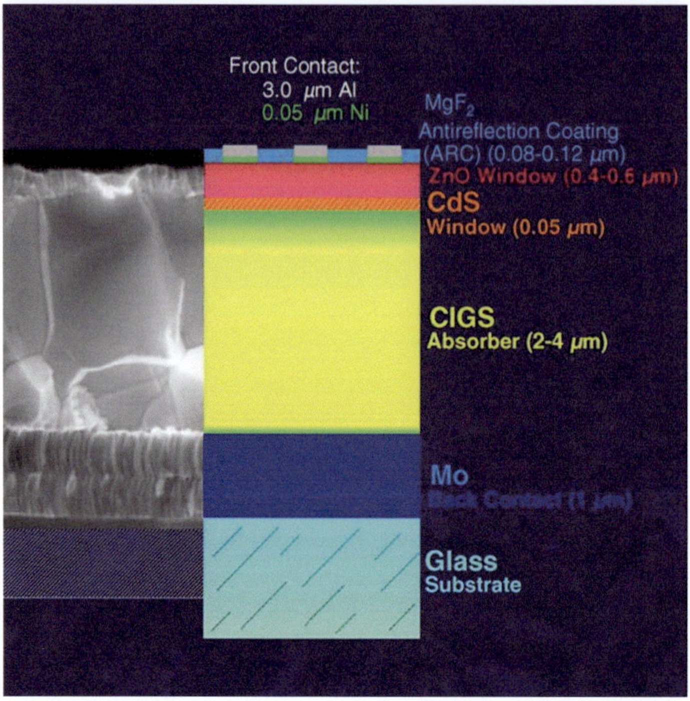

Fig. 1 Material stack up of CIGS solar cell

Processing CIGS in a Belt Furnace

The fabrication of CIGS solar cells within a production environment contains the deposition of copper, indium, and gallium on the selected substrate material while annealing them at elevated temperatures in controlled atmospheres. First, the substrate material is cleaned and then heated to an elevated temperature. Afterward, copper, indium, gallium, and selenium are deposited through a sputtering process and then the deposited parts are selenized in an elevated temperature profile. The selenization process involves ramping up from room temperature to 450 °C in about 4 min. The samples are soaked at this temperature for 7 min and then elevated to 550 °C for approximately 4 min. The samples are then held at 550 °C for another 7 min and then cooled down. Finally, while held at an elevated temperature, hydrogen sulfide gas is introduced to sulfurization. A detailed profile is illustrated in Fig. 2.

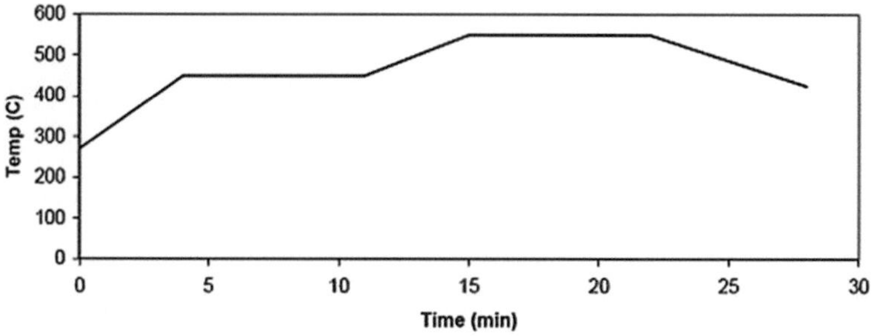

Fig. 2 Selenization profile of CIGS solar cell

Effect of Selenization Profile on the Efficiency of Solar Cell

The temperature and soaking times play two very important roles in determining the final efficiency of the CIGS solar cell. Many researches have revealed significant evidences for how the different sintering temperatures and soaking times have a direct impact on the efficiency outcome. The results of the experiment by Kadam et al. demonstrate that a sintering temperature above 500 °C and a soaking period between 30 and 60 min can improve the efficiency of the cell significantly.

In his experiments, when the samples were selenized at 400 °C for 10 min with a temperature ramp of 6 °C per minute, it resulted in smaller, non-faceted grains. This is an indication that either the temperature or the soaking time is inadequate. When the soaking time was increased from 10 to15 min, no significant improvements were noted in the results. Therefore, the ramp rates can be construed as having a larger impact on efficiency. In addition, Kadam et al. studied the impact of various heating rates. Changes in the parameters were made: heating rate was increased from 6 to 20 °C per minute and the temperature was increased to 425 °C with a soak period of 30 min. The resulting solar cell film appeared to be more homogeneous. Furthermore, the temperature was increased to see the effects on the grain size and cell efficiency, which indicated that at 475 °C, the grain size ranged from 0.5 to 1.5 um with an efficiency of 5.56% (Table 1).

At the same time, they detected that increasing the peak temperatures above 500 °C could also help with improving efficiency. However, at higher temperatures the grains become well faceted and less thick. It is fair to conclude that high-quality CIGS solar cells should be prepared at temperatures above 500 °C with an ideal soaking time between 30 and 60 min. Understanding that the heat rate, temperature, and soaking times are important factors in determining the ultimate cell efficiency, a belt furnace with a wide range of firing temperatures and a fast heat rate design are desired for constructing highly efficient solar cells.

Table 1 Effects of different sintering temperatures and soaking times

Condition temp, soak time	Grain size	Film thickness	Efficiency (%)	Comments
400 °C, 10 min	Varies from <1–2 um	5 um	<1	Inadequate parameters
475 °C, 30 min	0.5–1.5 um	3.5–3.6 um	5.56	Better efficiency obtained
500 °C, 30 min	2 um	2.7–2.8 um, rough surface	4.8	Nonuniform grain sizes
500 °C, 60 min	>1 um, well faceted	2.6 um	9	High efficiency
515 °C, 50 min	>1 um, highly faceted	2.6–2.7 um	8	High efficiency

Belt Furnace for CIGS Processing

A furnace that processes CIGS solar cells should be capable of operating to 650 °C or higher. The aim is for a wider belt with superior cross belt uniformity so that larger substrates can be contained in the future. Moreover, a muffle is required to ensure a cleaner operating environment because various processes in CIGS processing, such as sulfurization, involve the introduction of different gases at different time periods. The muffle needs to have the capability to control the gas type and gas flow in each zone. With all of this in mind, Torrey Hills Technologies has designed a furnace that is well suited to match the requirements for thin-film solar applications. Figure 4 illustrates a firing furnace that is capable of processing CIGS solar cells and Table 2 lists the technical specifications for the ideal CIGS furnace.

The designed HSA furnace uses ceramic heater boards in order to achieve elevated temperatures. Aside from the standard belt size of 350 and 650 mm, wider belts have been designed as well to accommodate wider glass substrates. While a 700 mm wide belt has been successfully engineered, efforts are underway to build a belt as wide as 1000 mm. As a standard feature, this furnace is equipped with a steel brush for cleaning the conveyor belt; however, ultrasonic belt cleaning is available as an extra option.

A microprocessor-based PID controller is what controls the furnace. Type K thermocouples are used in determining the zone temperatures, and the controls are located on the right-hand side of the furnace which can be viewed from the entrance. The central processing unit (CPU) is mounted under the exit table and the CPU is primed with a Windows operating system for easy computing. The computer system is preinstalled with a program for controlling the furnace parameters, including the belt speed and the zone temperatures. Temperature profiles can be stored and retrieved as well for future purposes. Thermocouple ports are located at the entrance table for connecting the profiling thermocouple directly into the microprocessor. This feature allows for the monitoring and recording of actual temperatures experienced by the part. Software is also included with the computer to capture, display, print out, and

Table 2 Specification of HSA 7503 belt furnace for CIGS processing

Specification	HSA 7503-08ZN
Rate temperature	900 °C max
	Normal operating: 850 °C
Belt width	30″ (750 mm)
Effective above belt clearance	1″ (25 mm)
Control zones	8
Conveyor speed	2–8 IPM (40–200 mm/min)
Loading table	59″ (1500 mm)
Unloading table	59″ (1500 mm)
Belt	Balanced V weave, SUS314
Heating elements	FEC heater board
Insulation	High-quality ceramic fiber
Temperature controller	Intelligent PID Shimaden controller
Alarm	Thermocouple, over temp, belt stop. Audio and visual alarm
Atmosphere	Six pipes of dry clean air or N_2. 2–6 m^3/h, 1.1–3.3 CFM
Cooling	Forced air cooling
Across belt temperature uniformity	+/− 4 °C
Overall system width	60.5′ (1540 mm)
Overall system length	343″ (8705 mm)
Overall system height	54″ (1350 mm)
Net weight	2500 kg
Power	Three phase, 480 VAC, 60 Hz, 38 KVA max
	Normal operating power draw is about 15 KVA

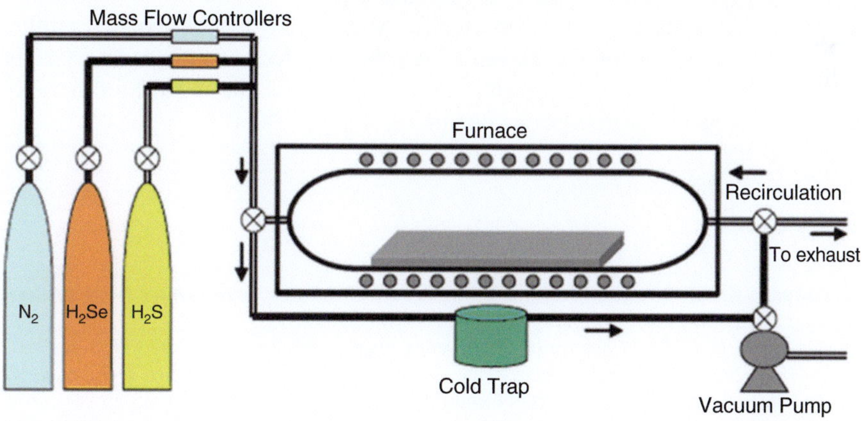

Fig. 3 CIGS sintering furnace requirement

store the furnace profile. Additionally, the furnace is equipped with a redundant over-heat safety protection system which incorporates an additional type "K" thermocouple in the center of each controlled zone and the multi-loop alarm. The specification of a HSA 7503 belt furnace is listed in Table 2 (Figs. 3 and 4).

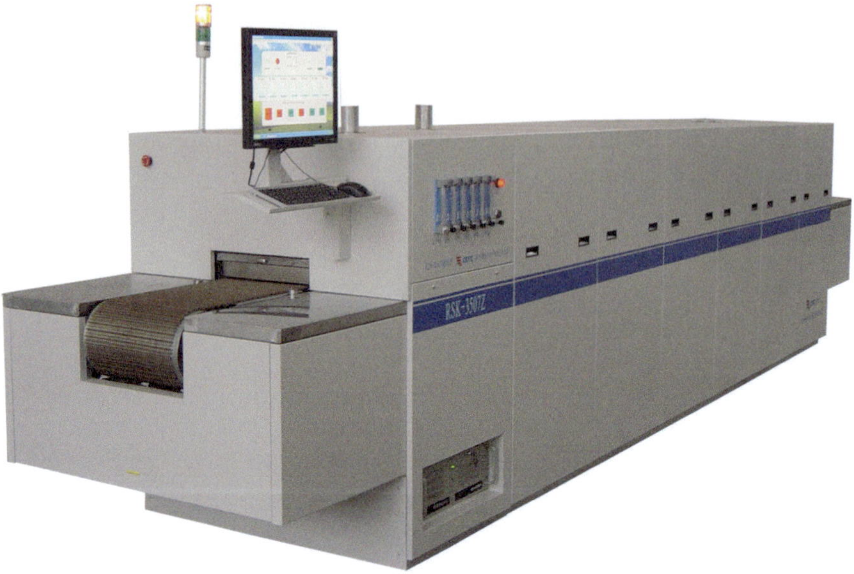

Fig. 4 HSA series belt furnace

Conclusion

The selenization process has a significant impact on the efficiency of CIGS solar cells. The proposed HSA furnace has the capabilities to elevate at a much higher temperature, maintaining cross belt uniformity and achieving the required profile. At the same time, the muffle design allows the introduction of specific gases in specific firing zones per requirement. In short, the HSA series is ideal for the selenization process in thin-film photovoltaics.

References

1. Edwards, B. M. (2008). *Screen and stencil print technologies for industrial N-Type silicon solar cells*. PhD thesis, University of New South Wales.
2. Kadam, A. A. (2006). *Preparation of efficient CuIn1-xGaxSe2-ySy/CDS thin-film solar cells by optimizing the molybdenum back contact and using diethylselenide as selenium precursor* (pp. 80–120). Orlando: University of Central Florida.
3. Kapur, V. K., Bansal, A., & Le, P. (2003). *Non-vacuum processing of CIGS solar cells on flexible polymeric substrates*. Inglewood: International Solar Electric Technology, (ISET).
4. Annual Technical Report. (2005). *Lab to large scale transition for non-vacuum thin film CIGS solar cells. NREL Subcontract No. XCQ-230630-30*. Inglewood: International Solar Electric Technology, (ISET).
5. Dwyer, D., Repins, I., Efstathiadis, H., & Haldar, P. (2009). *Deposition of CuInAlSe2 films using co-sputtered precursors and selenization*. Albany: College of Nanoscale Science and Engineering, University at Albany, State University of New York.

Understanding the Influence of Belt Furnace and Sintering Parameters on Efficiency of Dye-Sensitized Solar Cells

Introduction

Sintering of the electrode layer (composed of TiO_2 and TCO) is an essential step when determining the overall efficiency of the solar cell. The process consists of a two-stage sintering procedure where sintering of the layer occurs before and after dye implementation. Important parameters to note are the growth temperature, soaking time, and the heat rate, all of which will directly affect the outcome on the efficiency of the solar cell. Additionally, the materials selected for each segment of the cell will have an integral effect on the final efficiency as well. This paper will explore the ideal sintering process parameters and the material selections, as well as the impact they both play on the overall efficiency of the solar cell.

Structure

A dye-sensitized solar cell is composed primarily of three parts (as shown in Fig. 1). The first part, the substrate, is the negative terminal. The substrate has a layer of transparent glass on the outside and a coating of transparent conductive oxide (TCO) on the inside. This warrants sunlight to pass through. In the center sector, a layer of dye sensitizers binds to a layer of nanostructured titanium dioxide (TiO_2), where the TiO_2 is connected to the negative terminal to collect sunlight. All of the layers are then immersed in an electrolyte solution to allow charge transportation. The top part is the positive terminal and it contains a coating of carbon (graphite) or platinum for the purpose of transferring electrons. The outside layer is made of transparent glass and the top and bottom divisions are joined together to prevent the centered portion from leaking.

J. Xu et al., *Conveyor Belt Furnace Thermal Processing*,
https://doi.org/10.1007/978-3-319-69730-7_4

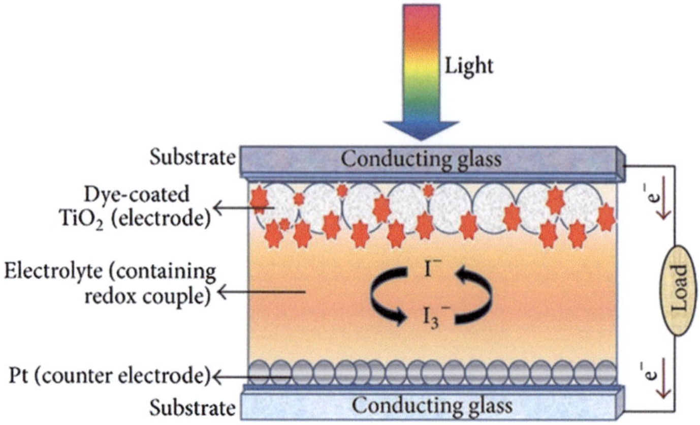

Fig. 1 Structure of a dye-sensitized solar cell

Electricity Generation Schematics

Initially, sunlight passes through the transparent conductive oxide (TCO) layer into the dye-sensitized layer exciting the electrons within the molecules. The electrons are then injected into the TiO_2 particles (which act as a semiconductor) transporting light-induced electrons toward the negative terminal. The negative terminal layer, or the TCO layer, is where all of the electrons are collected and then transported to the external circuit, generating electricity. Subsequently, the electrons are reintroduced into the solar cell through the positive terminal into the electrolyte. From here, the electrolyte transports the electrons back into the dye molecules, and the process is repeated (Fig. 2).

Material Selection

The transparent coating for the negative terminal is made of a thin layer of fluorine-doped tin oxide. It is a substrate that enables sunlight to pass through it, conducting electricity.

For the semiconductor, either zinc oxide (ZnO) or titanium dioxide (TiO_2) can be applied. TiO_2 is the preferred choice and it is used most often because the surface is highly resistant to continuous electron transfer. Titanium dioxide, however, is not sensitive to visible light and it will only absorb a small amount of solar photons. Therefore, dye sensitizers have to be joined together with the titanium dioxide layer in order to harvest large portions of the sun's light. Zinc oxide has higher electron mobility than titanium dioxide; however, it has a limited selection of organic dyes. This makes it a less suitable option until more exceptional alternative sensitizers are identified.

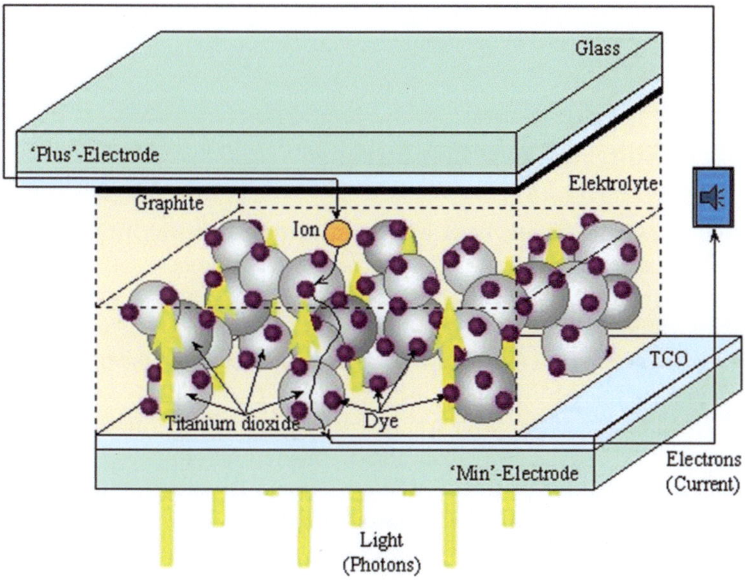

Fig. 2 Electricity generation schematics

Dye sensitizers can contain several classifications of materials. One option, the natural dye, can come from a variety of resources such as blueberries, blackberries, and raspberries. They are the easiest to come by and are excellent for student training courses and/or for testing purposes. A second option, the synthetic dye, provides better performance because of its optimized light collection property.

The material used for the positive terminal layer, or the cathode layer, can be platinum or carbon (graphite). Platinum is more efficient; however, the carbon alternative is the easiest and least expensive to use (great for school work or testing purposes).

Advantages

The advantage of DSSC compared to alternative options for solar electrical generation is that DSSC has a high price/performance ratio. This means that it is less expensive to assemble, as well as light and mechanically robust due to its material properties. With its ultrathin profile, the cells can be constructed into several shapes in order to meet the specific requirements for its design. Furthermore, due to its robustness, DSSCs maintain their efficiency at high temperatures. The efficiency of the DSSC is between 10% and 11%, which is higher than other thin-film solar panels having an efficiency of only about 5–13% on average. DSSC technology also operates in low-light conditions (e.g., cloudy weather where there is no direct sunlight), which is another reason why the DSSC is highly attractive among current choices for solar electrical generation.

Process of Sintering in a Belt Furnace

To start, the titanium dioxide layer is compounded by sintering TiO_2 nanoparticles at a temperature range of 300–500 °C. The sintering process takes place on the transparent conductive oxide (TCO) glass plate, which is put into a uniformly heated furnace for about 20 min. This process omits the ambient moisture within the titania layer, which is needed to ensure the electrical contact between the titanium dioxide nanoparticles and a worthy adhesion to the TCO (transparent conductive oxide) glass plate. Sintering of the layer can be executed at 150 °C; however, its performance power will be lower than those sintered at 450 °C (mentioned in P.M. Sommeling research paper; *ECN Solar Energy*). The layer is then soaked in the dye solution so that the dyes can be absorbed into the TiO_2 surfaces. Finally, the layer is inserted into a drying furnace where the titania is baked at 100 °C and then sanctioned for cooling.

Influence of Sintering Temperature on Performance of DSSCs

Extensive research has been accumulated on the effects of various temperatures and the efficiency of the DSSC. The current ideal firing temperature is preferred around 450–500 °C, as shown below. High sintering temperatures at 450 °C result in a more desirable contact between the nanoparticles and a stronger adhesion to the substrate than those sintered at lower temperatures. It is important to keep in mind, however, that the DSSC will become unstable at very high temperatures because they have an upper limit of 600–650 °C.

Sintered for 60 min

Temperature (°C)	Time (min)	ZnO efficiency (%)	TiO_2 efficiency (%)
100	60	0.55	
200	60	0.64	0.8
300	60	0.78	
400	60	1.06	1.29
500	60	0	

Below are alternative results discovered through experimental research.

Sintering time 60 min

Temperature (°C)	Time (min)	Average particle size (nm)	TiO_2 efficiency (%)
100	60	9.8	1.4
150	60	10.7	1.49
200	60	15.3	2.59
250	60	18.9	4.25
300	60	22.5	5.19

The temperature profile comparisons clearly indicate that higher sintering temperatures result in larger TiO_2 nanoparticles, which, in turn, result in an overall improvement on the efficiency of the solar cell. The reason being that the larger particles allow for prominent dye absorption, producing higher electron generation.

The sintering temperature limit

Temperature (°C)	Time (min)	Efficiency (%)
500	1	0.1
500	15	2.6
500	30	2.9
600	30	2.6
800	30	0.2

The temperature upper limit is around 600 °C. As you can see, the efficiency drops suddenly due to the instability of the TiO_2 nanoparticles.

In conclusion, a sintering temperature between 400 and 500 °C will result in a highly efficient DSSC.

Furnace for DSSC Application

The HSK Series furnace is an energy efficient precision thermal processing system specifically designed, and most often used, for DSSC applications. It has six-channel temperature profiling units for independent temperature profiling with an LCD data display and check, analysis software, sampling unit, 3 T.C., and an RS232 CPU interface. The HSK Series is designed to support continuous on/off heating and cooling cycles resulting from alternating production periods and inactive operation. The heating length of the HSK Series is 3220 mm (127″) and includes seven independently controlled heat zones. Process materials are transported through the furnace on a belt that is 350 mm (14″) in length with 50 mm (2″) of product clearance. The speed of the belt ranges from 40 to 200 mm (2–8″) per minute and is administered using a digitally displayed variable frequency motor controller. The belt speed is also programmable in IPM with readout right on the PC. The belt material on the HSK Series furnace is Nichrome V mesh (Balanced Spiral) and operates from 480 V, 3 phase, 5 wire, and 60 Hz with a maximum load connection of 42 kVA.

The performance of the HSK Series furnace is unparalleled as it can protect itself from overheating, overloading, and low gas pressure. It has an ultraclean low-mass refractory heating chamber that can increase heat from ambient temperature to 1050 °C in approximately 40 min. The temperature of the furnace is controlled by a microprocessor that typically operates from 200 to 900 °C. Each zone is managed using a high-performance, single ASIC full auto-tuning PID and a single loop intelligent temperature controller. The HSK Series atmosphere distribution and management system can terminate thermal shock and process contamination, as well as extract burn-off effluents across the entire width of the chamber for yield improvement. The HSK Series is assembled with entrance/exit curtains and an air powered Venturi exhauster (200 mm /8″ in diameter) to keep the firing chamber clean while, at the same time, improving temperature stability for drying and firing. The exhaust flow can also be easily adjusted using the flow meter. The HSK Series is readied with a redundant overheating safety protection system that incorporates a type "K" thermocouple (located in the center of each heated zone) and a multi-loop alarm. It ensures consistent "firing" results because of its exceptionally reliable temperature uniformity control. The HSK Series furnace has a removable condensate collection trap and provides emergency off buttons located at each end of the furnace (connected to a 24 V emergency off circuit). To see a complete list of the HSK Series specifications, please see the chart below.

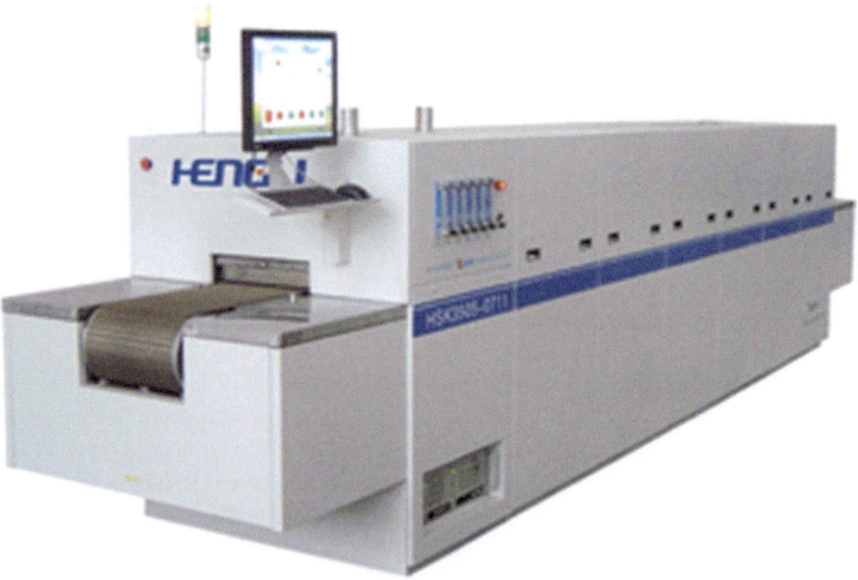

Specification	HSK2505-0611	HSK3505-0711	HSK6305-0711
Rate temperature	1050 °C	1050 °C	1050 °C
Belt width	250 mm/10″	350 mm/14″	635 mm/25″
Tunnel height	50 mm/2″	50 mm/2″	50 mm/2″
Heating length	2160 mm/85″	3220 mm/127″	3220 mm/127″
Cooling length	1200 mm/47″	1200 mm/47″	1200 mm/47″
Control zones	6	7	7
Conveyor speed	40–200 mm(2–8″)/min	40–200 mm(2–8″)/min	40–200 mm(2–8″)/min
Overall system width	1040 mm/41″	1100 mm/43″	1400 mm/55″
Overall system length	5800 mm/228″	7000 mm/276″	7000 mm/276″
Overall system height	1350 mm/53″	1350 mm/53″	1350 mm/53″
Typical temp. uniformity	+/− 2 °C	+/−2 °C	+/−3 °C
Net weight	1000 kg	1600 kg	2000 kg
Power	240 V, 3 phase, 60 HZ, 5 wire, 36 KVA	240 V, 3 phase, 60 HZ, 5 wire, 42 KVA	240 V, 3 phase, 60 HZ, 5 wire, 56 KVA
Spare part	1 set heating board, 1 solid relay, 2 relays, 2 switches	1 set heating board, 1 solid relay, 2 relays, 2 switches	1 set heating board,1 solid relay, 2 relays, 2 switches

Belt Furnace Parameters

Cleanliness, atmosphere, temperature controller, and sensors.

References

1. Bisquert, J. *Dye-sensitized solar cells*. From http://www.elp.uji.es/juan_home/research/solar_cells.htm
2. Wikipedia resources. (2011). From http://en.wikipedia.org/wiki/Dye-sensitized_solar_cell
3. Bowerman, B., & Fthenakis, V. (2001). *EH&S analysis of dye-sensitized photovoltaic solar cell production*. Upton: Brookhaven National Laboratory.
4. Watson, T., & Mabbett, I., et al. (2010). *Ultrafast near infrared sintering of TiO2 layers on metal substrates for dye-sensitized solar cells*. From online library at http://onlinelibrary.wiley.com/doi/10.1002/pip.1041/pdf
5. Pan, J. (2008). *Material property study on dye sensitized solar cells and CU (GA, IN) SE2 solar cells*. Oxford: Miami University.
6. Chou, T. P., Zhang, Q., Russo, B., Fryxell, G. E., & Cao, G. (2007). *Titania particle size effect on the overall performance of dye-sensitized solar cells*. University of Washington.
7. Bisquert, J., Cahen, D., Hodes, G., Rühle, S., & Zaban, A. (2004). Physical chemical principles of photovoltaic conversion with nanoparticulate, mesoporous dye-sensitized solar cells. *Journal of Physical Chemistry B, 108*, 8106–8118.
8. Sommeling, P. M., Spath, M., et al. (2000). *Flexible dye-sensitized Nanocrystalline* TiO_2 *solar cells*. Petten: ECN Solar Energy.

Manufacturing Solar Cells: Assembly and Packaging

Introduction

The trend in packaging has shown a serious shift: attendance at assembly and packaging conferences has been dwindling over the past few years. At the same time, solar power shows have been celebrated with lots of fanfare, aisles crowded. More and more electronics assembly and packaging companies are appearing at solar expos. The shift is on and there are significant opportunities for electronics engineers in the fast-growing solar business.

Solar cells grew out of the 1839 discovery of the photovoltaic effect by French physicist A. E. Becquerel. However, it was not until 1883 that the first solar cell was built by Charles Fritts, who coated the semiconductor selenium with an extremely thin layer of gold to form the junctions. The device was only about 1 % efficient. Subsequently Russian physicist Aleksandr Stoletov built the first solar cell based on the outer photoelectric effect (discovered by Heinrich Hertz earlier in 1887). Albert Einstein explained the photoelectric effect in 1905 for which he received the Nobel Prize in Physics in 1921. Russell Ohl, working on the series of advances that would lead to the transistor, developed and patented the junction semiconductor solar cell in 1946.

Today's solar cells can be described as the coexistence of three different generations: crystalline silicon, thin film, and dye sensitized. Along with the development of solar cells, there has also been a parallel development of solar cell manufacturing technologies. Assembly and packaging engineers have played a significant role in developing these manufacturing techniques, creating incredible potentials in every generation of the solar business.

J. Xu et al., *Conveyor Belt Furnace Thermal Processing*,
https://doi.org/10.1007/978-3-319-69730-7_5

First Generation

Elemental or crystalline silicon is the principal component of most semiconductor devices, most importantly integrated circuits or microchips. Silicon's ability to remain a semiconductor at higher temperatures has made it a highly attractive raw material for solar panels. Silicon's abundance, however, does not ease the challenges of harvesting and processing it into a usable material for microchips and silicon panels. At least three standard manufacturing processes mean that there are technical opportunities for assembly and packaging engineers.

Phosphorus Diffusion

There are two main layers that are essential to the solar cell's function. One is a p-type layer, which means that the wafers are boron doped, and an n-type layer created by introducing phosphorus. The silicon wafer usually already starts off by already being doped with boron. To form the n-type layer, phosphorus has to be introduced to the wafer at high temperatures of around 870 °C for 15–30 min in order for it to penetrate into the wafer. The excess n-type material is then chemically removed.

These diffusion processes are usually performed through the use of a batch tube furnace or an in-line continuous furnace. According to BTU, detailed cost of ownership models has shown that in-line diffusion can deliver per wafer costs to as low as one third the cost of a batch diffusion furnace. The basic furnace construction and process are very similar to the process steps used by packaging engineers.

Silicon Wafer Metallization

Electrical contacts are formed through squeezing a metal paste through mesh screens to create a metal grid. This metal paste (usually Ag or Al) needs to be dried so that subsequent layers can be screen printed using the same method. As a last step, the wafer is heated in a continuous firing furnace at temperatures ranging from 780 to 900 °C. This completes the metallization process, removes solvent and binder, and forms electrical contacts. Metallization is the most critical step. The challenge of reducing wafer thickness for higher efficiency has created stringent requirements for both the equipment and the process itself (Fig. 1).

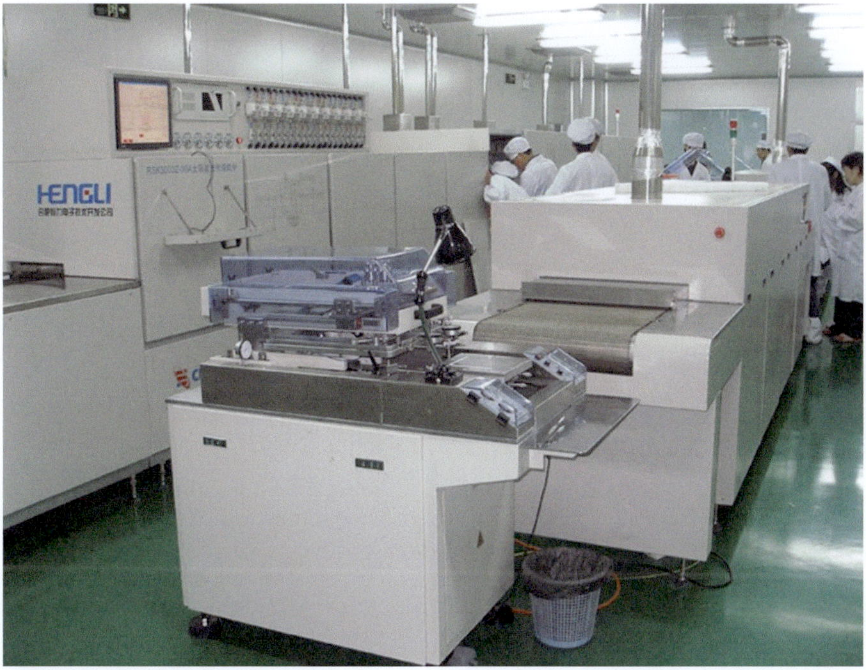

Fig. 1 Si wafer metallization line

Solar Module Assembly

Solar module assembly usually involves soldering cells together to produce a 36-cell string (or longer) and laminating it between toughened glass on the top and a polymeric backing sheet on the bottom. Frames are usually applied to allow for mounting in the field, or the laminates may be separately integrated into a mounting system for a specific application such as integration into a building. The basic process is very similar to the SMT process assembly that packaging engineers are already familiar with, albeit on a larger scale. The packaging industry's lean manufacturing methodology can be applied directly to solar module assembly.

Second Generation

Second-generation solar cell, also known as thin-film solar cell (TFSC) or thin-film photovoltaic cell (TFPV), is made by depositing one or more thin layers (thin films) of photovoltaic material on a substrate. The most advanced second-generation thin-film materials in use today are amorphous silicon (aSi), cadmium telluride (CdTe), and copper indium gallium selenide (CIGS). The thickness range of such a layer is

Fig. 2 Thin-film solar cell annealing furnace

wide and varies from a few nanometers to tens of micrometers. Is thin film now the way to go? There are certainly many good reasons for moving to thin films for the solar cell manufacturing process.

Thin-Film Deposition

Copper indium gallium selenide (CIGS) is used for the thin-film active layers in CIGS solar cells, commonly formed using sputter deposition. During this vacuum-based process, a plasma of electrons and ions is created from inert argon gas. These ions dislodge atoms from the surface of a crystalline material which is then deposited to form an extremely thin coating on a substrate. Depositing thin film by sputtering is the same process used in semiconductor manufacture and in packaging.

Thin-Film Annealing

After sputtering, the thin film needs to be annealed to achieve optimum results. It is also possible to inject additional chemicals during the annealing process. An annealing furnace is similar to the brazing furnace commonly used in packaging industries as shown in Fig. 2. The muffle is typically made of SUS 316L material to ensure good corrosion resistance for the thin-film solar panel's corrosive environment. A typical belt furnace can anneal up to 600×1200 mm (23.6×47.2 in.) thin-film solar panels after thin-film deposition.

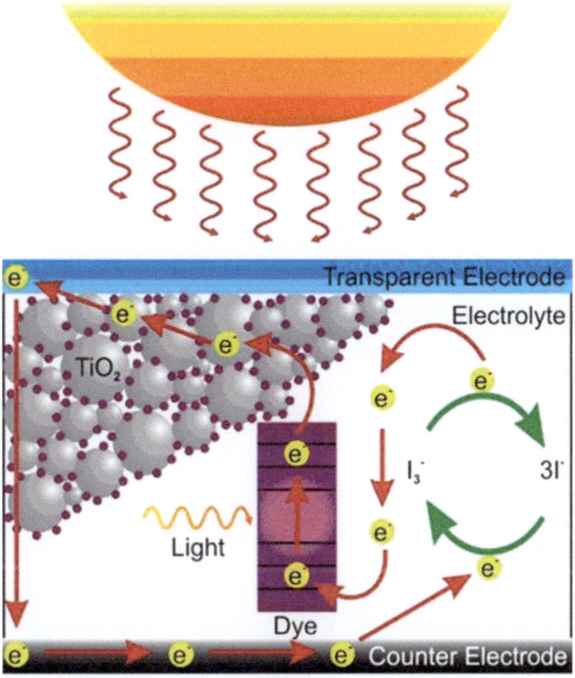

Fig. 3 Schematic diagram for dye-sensitized solar cell

Metallization

Like its first-generation cousin, the manufacture of thin-film solar cells needs Al or Ag screen-printing metallization, originally invented for the thick film process. Such metallization pastes or inks can be used on both rigid (glass, silicon) and flexible (polyimide, polyester, stainless steel) substrates. The metallization can be accomplished through either thermal curing or firing.

Third Generation

The electrochemical dye solar cell was invented in 1988 by Professor Graetzel of Lausanne Polytechnique, in Switzerland. The "Graetzel" dye cell uses dye molecules adsorbed onto the nanocrystalline oxide semiconductors such as TiO_2 to collect sunlight. Dye cells employ relatively inexpensive materials including glass, titania powder, and carbon powder (Fig. 3).

Graetzel's cell is composed of a porous layer of titanium dioxide nanoparticles, covered with a molecular dye that absorbs sunlight, like the chlorophyll does in green leaves. The titanium dioxide is immersed in an electrolyte solution, above which is a platinum-based catalyst. As in a conventional alkaline battery, an anode

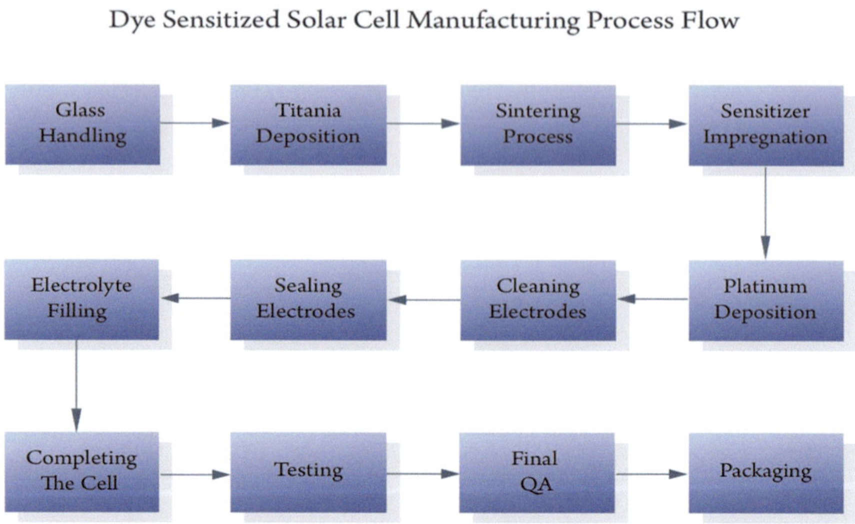

Fig. 4 Manufacturing process of dye-sensitized solar cells

(the titanium dioxide) and a cathode (the platinum) are placed on either side of a liquid conductor (the electrolyte). Sunlight passes through the cathode and the conductor and then withdraws electrons from the anode, at the bottom of the cell. These electrons travel through a wire from the anode to the cathode, creating an electrical current (Fig. 4).

Similar to Semiconductor

The basic dye cell manufacturing steps also resemble the approaches taken by the semiconductor and packaging industry. For example, a screen printer is typically used to apply titania and other layers to the transparent conductive optical (TCg or TCO) glass. Nanocrystalline TiO_2 pastes are screen printed onto the TCO glass, then dried and fired in a continuous belt furnace. The sintering process allows the titanium dioxide nanocrystals to partially "melt" together, in order to ensure electrical contact and mechanical adhesion on the glass. All these furnaces are typically modified from standard thick film furnaces.

After dye staining and anode-side application of proprietary current collectors, platinum catalyst is obtained by using the Pt-catalyst T/SP product which can either be squeegee printed or screen printed using a polyester mesh of 90. The solar cell needs to be dried at 100 °C for 10 min before being fired at 400 °C for 30 min. During the assembly, sealing, and filling processes, TCO glass with the completed titania layer is mated to the cathode current collector, protective glass plate sealed, bus bar attached to the cell, and then the cell is filled with electrolyte. Custom designed, fully automated, and efficient cell assembly, sealing, and electrolyte filling machine sets are required for these production steps.

At one time, Torrey Hills Technologies sold in-line continuous furnaces mostly for thick film and brazing applications. Several years ago, in response to the growing demands of the solar manufacturing industry, the company's engineers reinvented the original technology and adjusted it to different types of solar cell processing. A critical step in solar cell manufacturing is metallization through screen printing. By changing the specifications of thick film drying and firing furnaces, the company stepped comfortably into the solar cell market.

Conclusion

Solar technologies have created compelling technical challenges and business opportunities for assembly and packaging engineers. The traditional thick film, thermal treatment, and assembly techniques play key roles in solar cell manufacturing. Many skill sets possessed by electronics engineers can be easily reinvented and applied to the solar cell industry.

A Practical Guide for Improving Crystalline Solar Cell Efficiencies Through Optimization of the Firing Process

Bjorn Dahle

Introduction

The crystalline solar cell manufacturing industry has long acknowledged that when solar wafers are processed with an optimal thermal profile, the solar cell efficiency increases. Studies from Heraeus and KIC have quantified that such process optimization may lead to significant increases in cell efficiencies up to several tenths of percentage points. (Refer to the end of the article for a link to a Heraeus and KIC study.[1]) Because the solar cells are priced as a function of their efficiencies, the thermal process optimization potentially represents a significantly profitable endeavor.

The difficulty, however, is twofold: (1) to obtain accurate and repeatable profile readings, and (2) to identify the correct furnace recipe to achieve an optimum thermal profile.

This document focuses on the use of KIC's new solar products (e-Clipse TC attachment fixture, SunKIC profiler, Spectrum process optimization software) to help silicon solar cell manufacturers improve cell efficiencies while maintaining or improving productivity. The examples used come from the metallization process; however, the principals presented also are applicable for other thermal processes.

Thermal Process Development

As opposed to more mature industries such as semiconductor and electronics assembly, the solar industry currently does not have a clear understanding of the ideal wafer profile or process window for each unique application (wafer, silver paste,

[1] Link to the Heraeus-KIC study on increased cell efficiency as a result of profile optimization: http://globalsolartechnology.com/index.php?option=com_content&task=view&id=4038&Itemid=5

J. Xu et al., *Conveyor Belt Furnace Thermal Processing*,
https://doi.org/10.1007/978-3-319-69730-7_6

furnace and other variables). There may be numerous wafer properties and variables in the processes upstream from the metallization furnace that affect the ideal wafer profile. Therefore, each manufacturer must perform a design of experiment (DOE) to identify the best wafer profile or range of profiles (process window). Such a DOE typically involves changing the wafer profile while the cell efficiency, fill factor, and other quality measurements are measured. The profile is changed numerous times (on identical wafers that have been processed up to the firing furnace) in a "trial and error" approach until the responsible engineer is satisfied with the cell efficiency.

The wafer profile itself is a result of the furnace settings and how the thermodynamic properties of the furnace heat and cool the wafer. A modern metallization (firing) furnace can be set up using tens of millions of alternative recipes (combination of zone temperatures and conveyor speed), making the DOE both difficult and time consuming. Even worse, when changing one profile parameter, e.g., the peak temperature, all the other parameters such as time above 500 °C, time above 600 °C, ramp rate, etc. also change. As a result, it becomes difficult to determine what caused the improvement in the cell efficiency.

Finally, the traditional methods to record the wafer profile suffer from inaccurate and non-repeatable measurements. This problem is caused by the method used to attach the thermocouples (TC) to the wafer. It is not uncommon to get a 50 °C difference in peak temperature readings from one profile to the next when taken only minutes later. Clearly, a new TC attach method is required before any purposeful DOE can take place.

Trial and Error Experimentation for Improved Wafer Profiles

The following steps are recommended to explore new and improved wafer profiles that result in higher cell efficiencies:

1. KIC's TC attachment method

In order to significantly improve the profile measurements, KIC developed a four-TC attachment fixture called the e-Clipse. The user simply slides his or her own wafer into the fixture and four spring-loaded TCs with flattened disk beads automatically seat themselves. The e-Clipse is both highly accurate and repeatable (Fig. 1).

2. Initial furnace setup and associated wafer profile

The starting point for the DOE is whatever furnace set points (recipe) that are currently being used. Alternatively, the responsible engineer may use his knowledge and experience to select the initial setup. One option for factories with multiple production lines is to use the profile in the furnace that yields the highest cell efficiency.

When selecting a particular combination of conveyor speed and zone temperatures, the furnace recipe will yield a specific wafer profile and associated cell efficiency as depicted in Fig. 2.

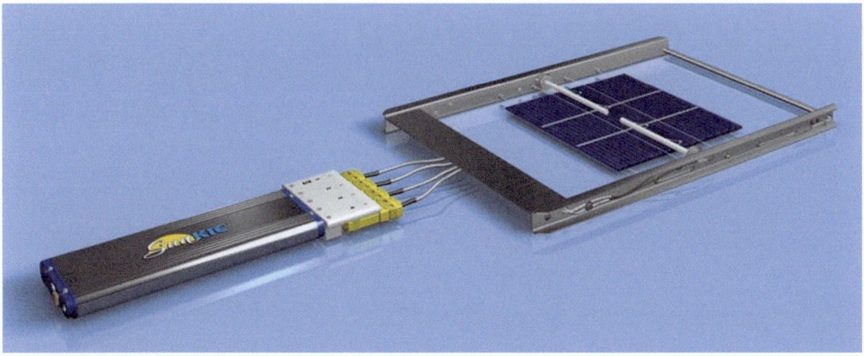

Fig. 1 The e-Clipse TC attachment fixture and SunKIC profiler

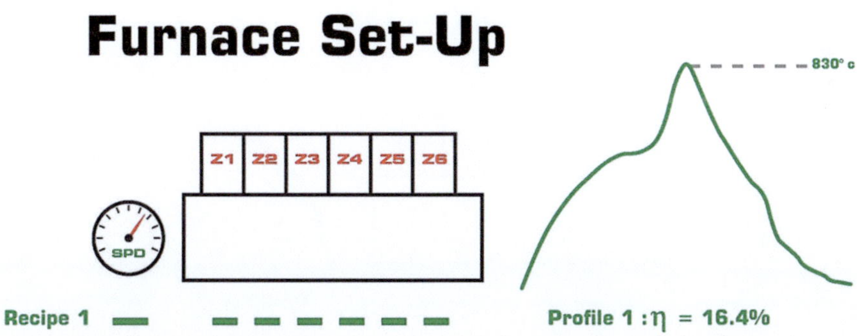

Fig. 2 Initial furnace setup and associated profile and cell efficiency

3. Second furnace recipe

To experiment with new and improved profiles, the responsible engineer may use his or her knowledge and information about wafer, silver paste properties, etc. to change the profile in certain directions. (The profile data and cell efficiency numbers in this article are only established to illustrate the process optimization method. The actual numbers vary with each application. The temperatures used in these examples are higher than what the industry has been used to seeing. This is a result of the more accurate SunKIC/e-Clipse readings that record data much closer to the actual wafer temperatures.)

In this example of a multi-crystal wafer, let us change the peak temperature in increments of 10 °C. We need to find the furnace recipe that increases the peak wafer temperature from the current 830 to 840 °C as seen in Fig. 3.

The engineer will set up the furnace to the newly chosen recipe, wait for the furnace to stabilize at the new temperature(s), run one or more wafers through the furnace, and measure their average cell efficiency.

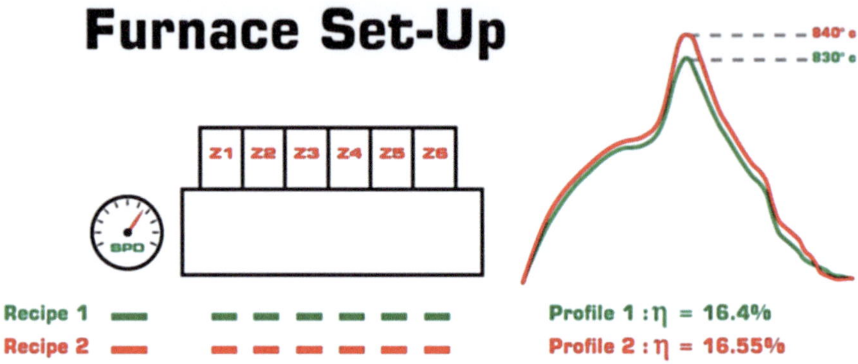

Fig. 3 New furnace recipe and associated wafer profile and cell efficiency

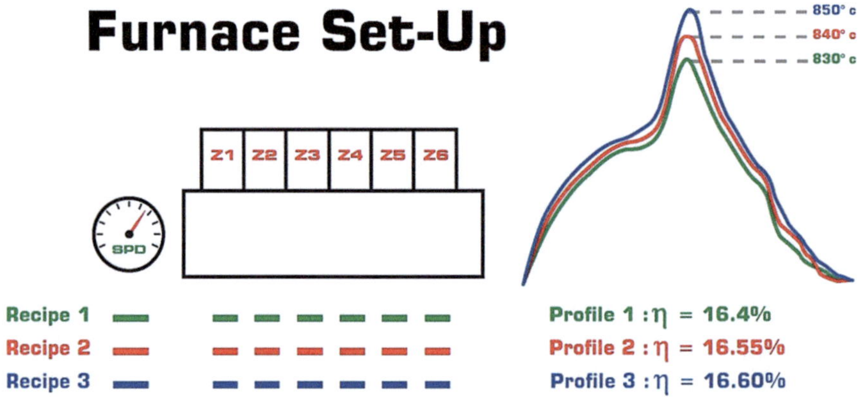

Fig. 4 Third recipe resulting in a profile yielding higher efficiency again

4. Third furnace recipe

The cell efficiency increased when running the furnace at recipe 2, so let us change the recipe again to achieve a peak temperature of 850 °C (Fig. 4).

5. Fourth furnace recipe

The cell efficiency improved (Fig. 4), so let us increase the peak temperature yet again, this time to 860 °C.

6. Fifth furnace recipe

The efficiency dropped (Fig. 5). It is tempting to conclude that the previous profile with a peak temperature of 850 °C is the best. The problem with this experiment is that although we focused on changing the peak temperature, all the other dimensions of the profile likely changed as well. Figure 6 shows how time above 500 °C and 600 °C kept changing. At this point, we do not know whether the peak temperature or time above 500 °C or 600 °C contributed to the changes in cell efficiency.

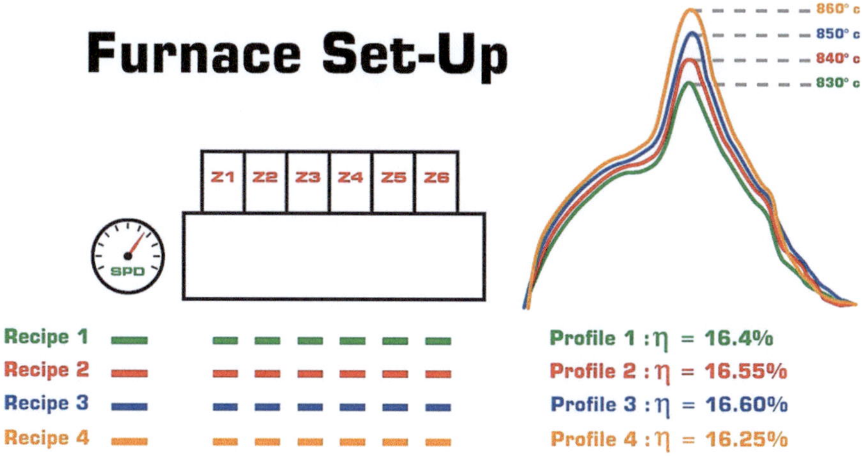

Fig. 5 New recipe resulting in a profile yielding lower cell efficiency

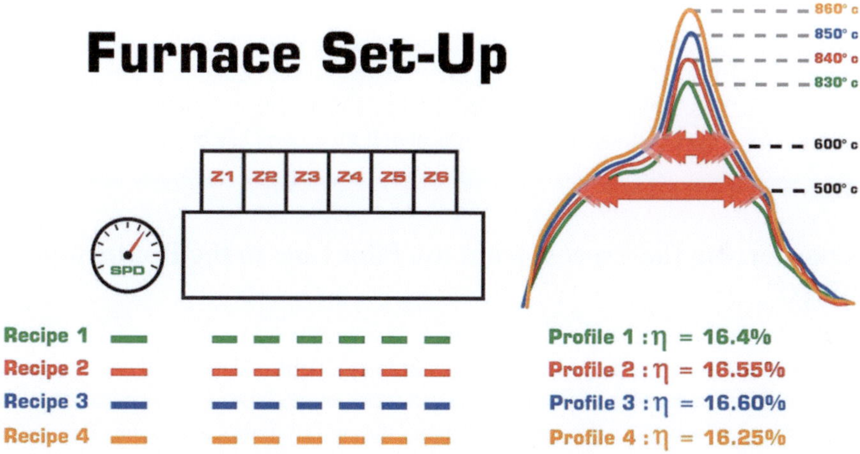

Fig. 6 DOE where multiple profile variables changed simultaneously

7. Select furnace recipes that only change the peak profile while keeping everything else constant

What is needed is the capability to only change one variable at the time. The KIC Spectrum enables the user to lock in all but one parameter at a time. The engineer needs to ask the software to identify the recipes that change the peak temperature in 10 °C increments while keeping time above 500 °C and time above 600 °C the same as depicted in Fig. 7. The Spectrum essentially evaluates all of the millions of alternative furnace recipes, and it predicts the resulting wafer profile for each. It then selects the furnace recipe that produces the requested profile. The Spectrum will provide its recommendations within seconds of completing the profile run.

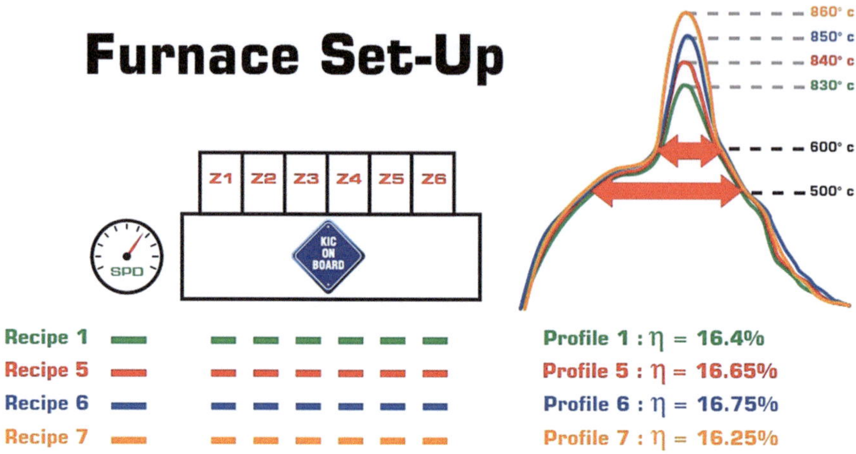

Fig. 7 New recipes increase the peak temperature on the wafer in 10 °C increments while keeping the other profile dimensions stable

The cell efficiency changes are now a direct result of the peak temperature only. Once the highest efficiency has been found based on the peak temperature (recipe 6 in Fig. 7), this dimension can be locked while changing one of the other profile parameters. This continues until the highest cell efficiency has been found.

Transferring the Process from the Pilot Line to the Production Lines

The simple but efficient process improvement guideline above will yield results quickly. Most companies will perform such work in the laboratory furnace or pilot line furnace. The next challenge is to transfer the optimum process (or profile) to the production lines (Fig. 8). Because the production line furnaces are different than the furnace where the process was developed, the process transfer may be difficult. Even if the production line furnaces are of the same make and model, they will have different thermodynamic properties due to wear and tear, aging heating lamps, preventive maintenance changes, and more.

The solution is to enter the optimal profile into the SunKIC profiler, run a profile, and to ask the Spectrum software to identify which setup will yield the requested profile for each production furnace (Figs. 9 and 10). Each production furnace will have different settings, but they all will produce the same wafer profile. The Spectrum software makes the process transfer extremely quick even when the furnaces are all different.

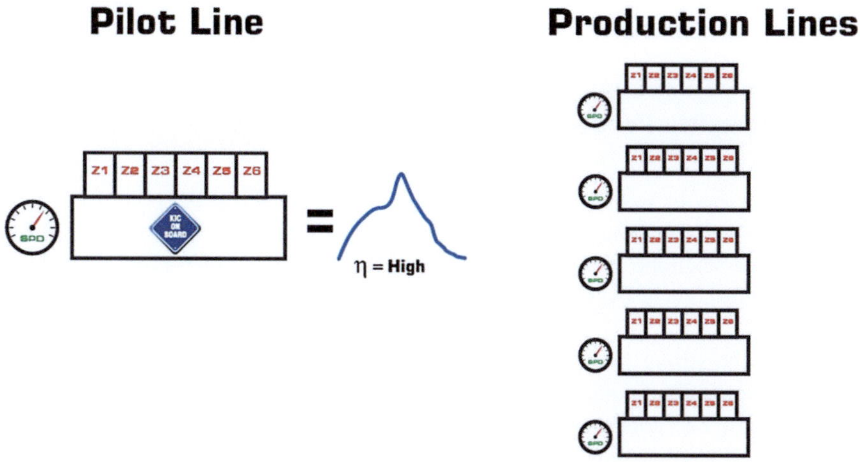

Fig. 8 How to transfer the optimized process to the production lines

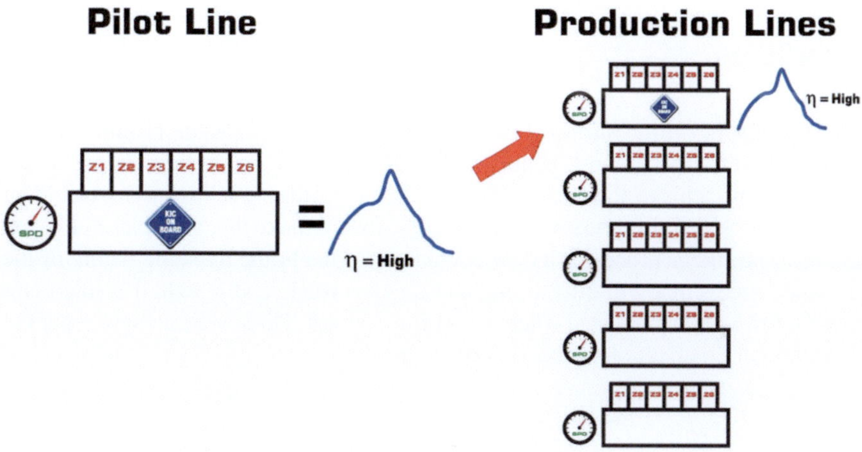

Fig. 9 Transfer the optimized process to the first production line

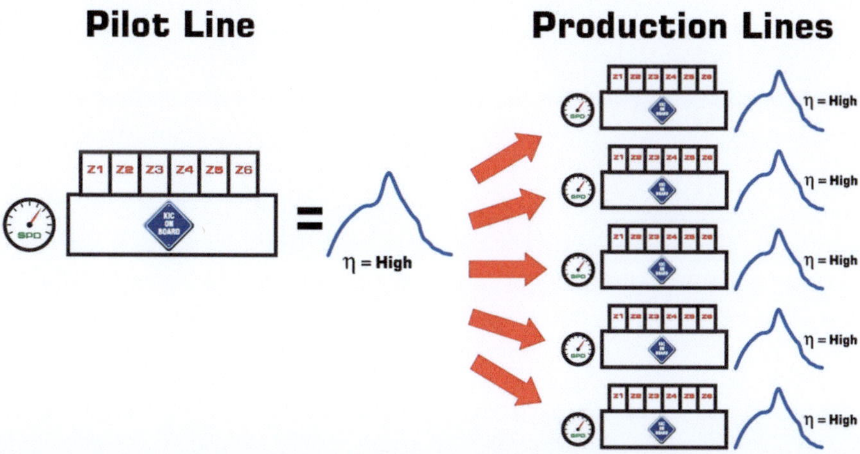

Fig. 10 Transfer the optimized process to all the production lines

Periodic Adjustment of the Production Furnaces to Consistently Produce Higher Efficiency Cells

The final aspect of setting up and running a solar cell firing process in the "sweet spot" of the cell efficiency is to periodically adjust the furnaces in a similar fashion to the above when the process starts drifting. The thermodynamic properties of the furnaces will change over time due to aging lamps, preventive maintenance, and more.

Due to the strong financial benefits of consistently producing solar cells with high efficiency, the furnaces should to be adjusted when the measured cell efficiency drops. The good news here is that the KIC Spectrum process optimization software will identify the correct furnace settings within seconds of running a profile, hence minimizing production downtime.

Conclusion

Fine-tuning the firing furnace to achieve a more optimal wafer profile leads to higher cell efficiencies and, hence, higher profitability. This article demonstrates easy steps to achieve this. KIC's SunKIC solar profiler, e-Clipse TC attachment fixture, and Spectrum process optimization software make this possible even when the ideal profile or process window is unknown. Due to the dynamic nature of the thermal process that will drift throughout the day, week, and month, it is important that the furnaces are adjusted when the profile changes. This can be achieved quickly with little or no effect on production downtime.

Study on the Effect of the Firing Process to the Energy Conversion Efficiency of Solar Cells

Zhiqing Sun and Jing Han

Background

Over the last few years, solar cell manufacturers have strived to improve energy conversion efficiency at a lower cost. Optimizing the metal electrode firing process in furnaces is a common way to achieve this goal. This was overlooked in the past. Most of the time, such thermal process work was done mainly by experienced engineers without much in-depth engineering study and development. This article will introduce a method of profiling and optimizing furnace set points in the crystalline silicon solar cell manufacturing process using SunKIC, in order to achieve optimal results for a stable and high solar energy conversion at Topsola's factory in Suzhou, China.

Principle of Experiment

After selecting the initial furnace set points, the actual profile on the surface of the crystalline silicon solar cell was measured and recorded using KIC's SunKIC profiler. By methodically selecting new furnace set points and analyzing the resulting profile characteristics and their respective energy conversion efficiency, the process window and optimal profile can be determined using the profiler software. The energy conversion efficiency of each cell will be calculated.

Material and Equipment List

Silicon solar cell: 125 × 125 mm
Aluminum paste: all samples use the same type of aluminum paste

J. Xu et al., *Conveyor Belt Furnace Thermal Processing*,
https://doi.org/10.1007/978-3-319-69730-7_7

Silver paste: front side Heraeus CL80-9235HL; rear side Heraeus SOL230S
Printer: ASYS
Furnace: Despatch CF7210
Profiler: KIC SunKIC

Experiment

A reference or base profile was selected to fire 30 pcs of solar cells. The average efficiencies were then computed. The final heating zone of the furnace was varied manually to develop different profiles. For each unique profile, 30 cells were fired. Table 1 shows the results of six different profiles. The seventh profile was determined by KIC's Spectrum optimization software. Figure 1 depicts the average cell efficiency for each set of 30 cells fired at the listed furnace set points.

The actual peak measured temperature also is displayed.

Analysis and Discussion

The box chart shown in Fig. 2 below depicts the results of the average energy conversion efficiency of solar cells fired at different profiles. We can see that result using profile 3, 4, 5, and 7 had a higher efficiency and less variation. From Table 1, we can see that the average peak temperatures were 842–865 °C. This is the peak temperature process window for firing such paste.

The optimized profile 7 in Fig. 3 has a 0.15% higher efficiency than the initial profiles. It also has better stability as compared to profiles 3, 4, 5, and 6. This reduces the number of lower efficiency cell produced, hence improving the productivity and cost at Topsola's factory in Suzhou, China.

Table 1 Average energy conversion efficiency of various profiles

No.	Avg. conversion efficiency %	Furnace setting °C 1	2	3	4	5	6	Actual peak temp °C
1	17.1276	400	500	610	700	800	895	820
2	17.1875	400	500	610	700	800	905	829
3	17.2766	400	500	610	700	800	920	842
4	17.2787	400	500	610	700	800	930	851
5	17.2778	400	500	610	700	800	945	865
6	17.1283	400	500	610	700	800	955	871
7[a]	17.2753	400	500	610	700	844	923	851

[a]Profile by KIC optimization software

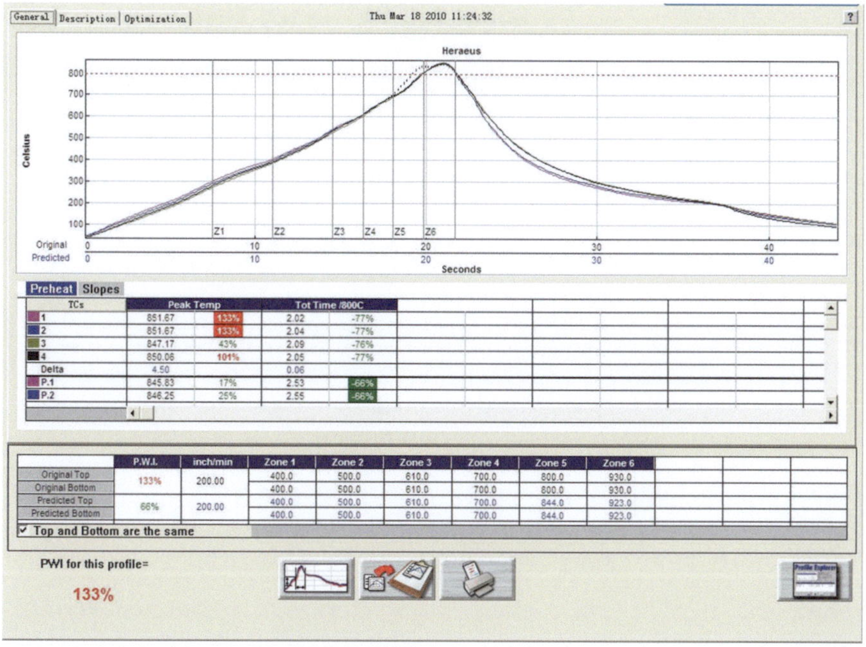

Fig. 1 Spectrum optimizing software interface

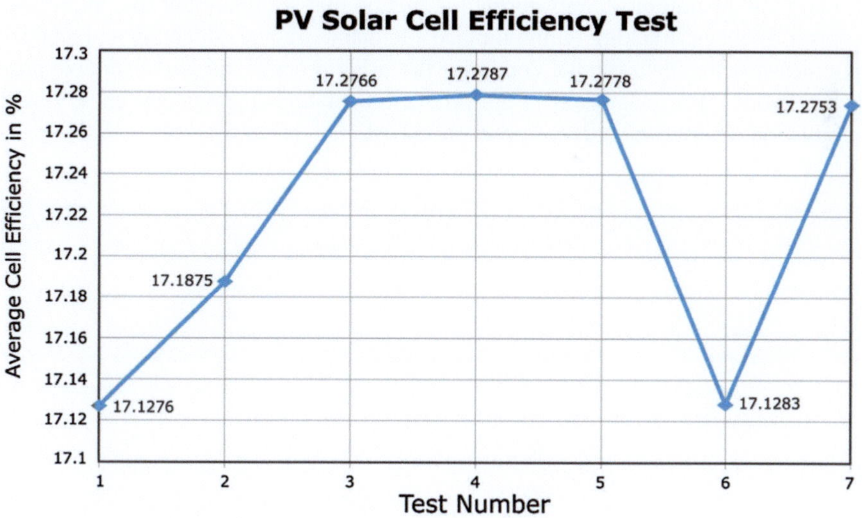

Fig. 2 Average energy conversion efficiency for various profiles

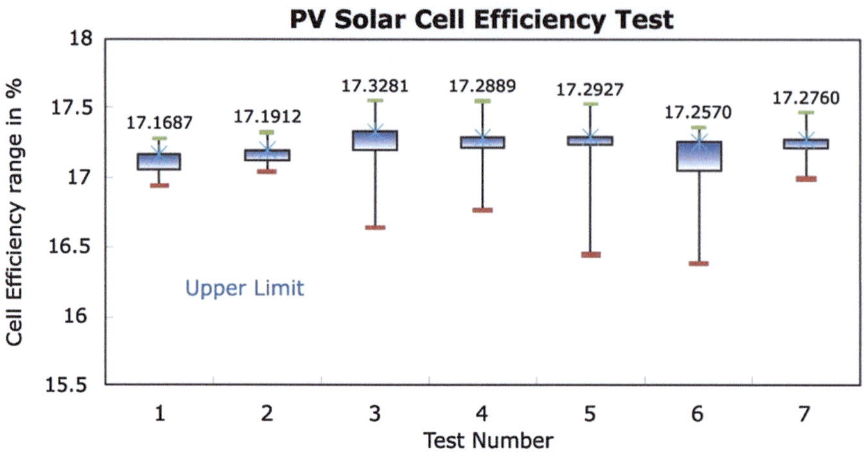

Fig. 3 Distribution chart of energy conversion efficiency for various profile

Conclusion

In summary, the accurate measurement of the cell profiles and optimization of the furnace's temperature settings during production play an important role in process control. Frequent tracking on the furnace's temperature and collecting and analyzing measurement data can help optimize the firing process and find a suitable process window. The end result is the production of higher efficiency cells, stable production quality, and reduced production cost.

Thermal Profiling of Silicon Solar Cells During the Metallization Process

Casey Kazmierowicz, Bjorn Dahle, Umesh Kumar, and Ed Graddy

Introduction

The metallization process is an important step in silicon solar cell manufacturing. During this step, maintaining the proper thermal profile of the wafer is very critical for high yield manufacturing of cells with good electrical performance. Thermocouples are commonly used to record the wafer thermal profile during the fast infrared (IR) belt furnace process. At present, two different methods are used to attach the thermocouples to the wafer. In the first method, the thermocouple beads are pinned to the surface with a spring load. In this method, placement and contact inaccuracy introduce errors. In the second method, thermocouples are cemented to the wafer surface. While it improves the contact, thermal properties of the cement greatly influence the readout. As a variation, highly polished single crystal wafers with carefully cemented thermocouple beads are used as reference wafers. These wafers allow the users to maintain and correct the furnace profiles as a function of time. Yet, it does not allow the user to measure the surface thermal profiles of the wafers being processed, which may differ considerably from that of reference wafers.

Cementing the thermocouple to the wafer can be avoided with the e-Clipse, a new frame design from KIC (Fig. 1). The e-Clipse can accommodate wafers up to 160 × 160 mm size.

It is very difficult to measure the wafer surface temperature in an IR belt furnace. When thermocouples are used for the measurements, the recorded values differ from the actual temperature noticeably. In addition to the thermocouple placement procedure, two additional factors, namely, thermocouple time constant and thermal contact resistance, affect the measurement accuracy. In the e-Clipse, low mass thermocouple junctions are flattened to increase the contact area to minimize thermal contact resistance and to improve sensitivity. In the IR heating environment, thermal contact resistance is also affected by the radiation intensity and the actual wafer

J. Xu et al., *Conveyor Belt Furnace Thermal Processing*,
https://doi.org/10.1007/978-3-319-69730-7_8

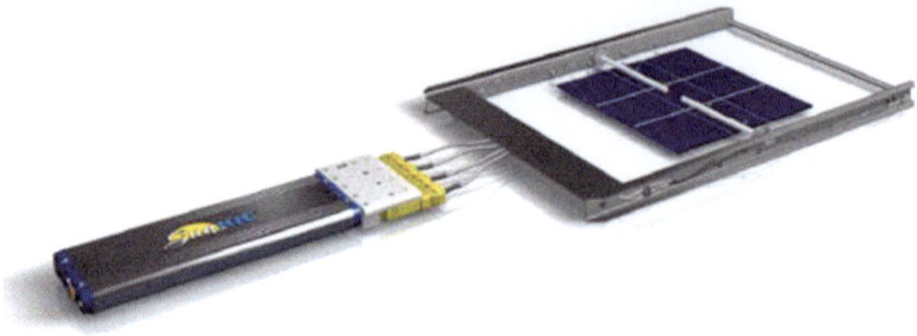

Fig. 1 e-Clipse with SunKIC datalogger

temperature. It is generally well known that measurement accuracy is valid only over a narrow temperature range.

In this paper, a procedure used to identify optimum TC junction geometry is described. Measurement accuracy and repeatability of the commercial units are also discussed.

Thermocouple Junction Design

As mentioned, in the rapid heating environment of an IR belt furnace, the temperature readings are greatly influenced by thermal contact resistance. Primary factors affecting the thermal contact resistance are contact area between the TC and the wafer, input power to the IR lamp, and the peak temperature.

To optimize the TC junction design, all of these factors were considered. At present, silicon solar cells are heated to approximately 800 °C during the metallization step. The soak time at the peak is maintained below 5 s.

To identify optimum design, test units with various shapes of TC junctions were made. In the test units, the contact surface areas of the thermocouples were varied from approximately 0.1 to 16 mm^2. In Fig. 2, one such test unit with two different junction shapes is shown.

To identify the actual wafer temperature, the melting phenomena of molybdenum trioxide (M003) was utilized. This compound melts at 795 °C. It does not react with the SiNx passivation layer which is on the emitter side of the silicon wafer.

Test Conditions

The surface temperature of a 6×6 in. bare poly silicon wafer was recorded with the test units. A small spot ($d < 3$ mm and $t \sim 10$ m) of M003 was applied on the wafer near one of the thermocouples. The measurements were carried out in a Despatch IR belt oven model number CF7210. The zone set points for the six-zone furnace

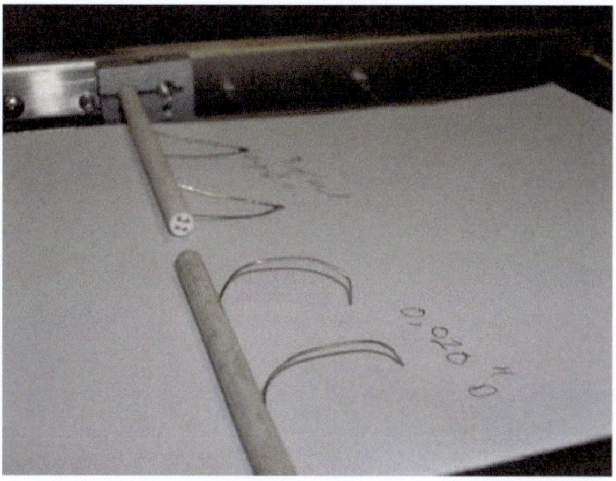

Fig. 2 An example of test unit with 0.02 in. mineral insulated sheathed type K thermocouples from Omega (bottom right) and flattened thermocouples (top left)

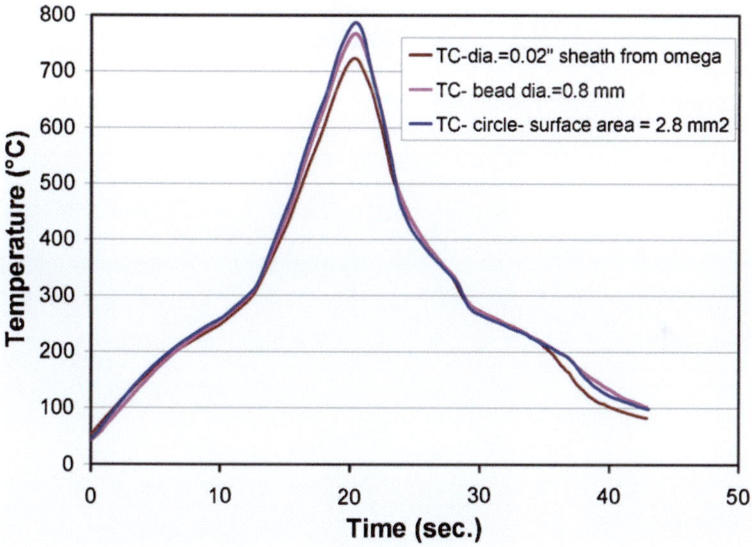

Fig. 3 Comparison of three different TC junction types run on the same wafer, through the same furnace, set to the same set point. The set point temperatures for zone 1 through zone 6 were 400-400-500-700-800-880. The belt speed was 200 ipm

were 400-400-500-700-800-880, and the belt speed was 200 ipm. These set points were selected to reflect the rapid firing conditions used in commercial scale during metallization process. Initially, the set point for the last zone was varied to identify the minimum required setting for M003 melting. Molybdenum oxide melting was verified with five repeats at these set points. Five additional measurements were conducted by reducing the set point of Z6 to 870 °C. In all cases, M003 did not melt.

In Fig. 3, the recorded profiles with three different thermocouple contact areas are compared.

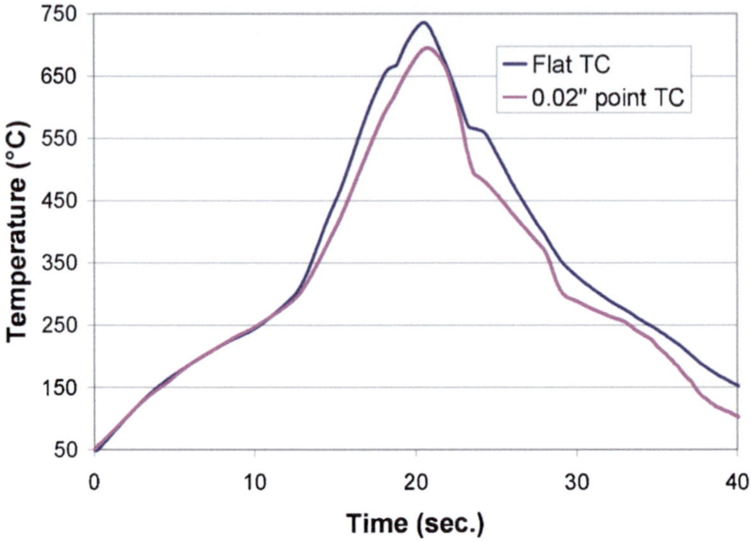

Fig. 4 Effect of TC junction shape on sensitivity

Fig. 5 TC junctions in commercial unit

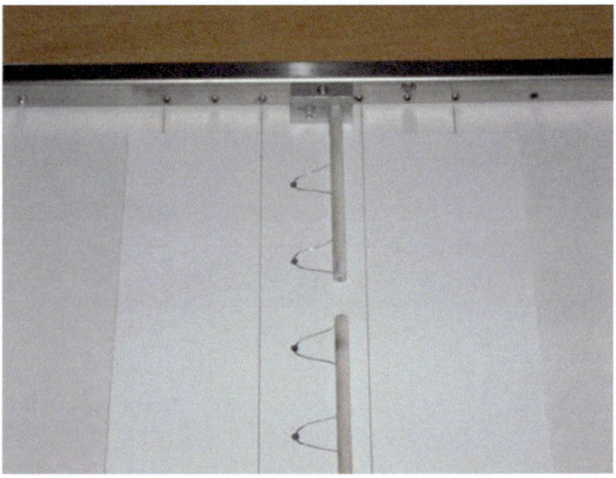

When the TC junction area was >3 mm^2, in general, the recorded peak temperature was closer to the actual wafer temperature.

In Fig. 4, the recorded profiles of silicon wafer with back aluminum paste registered with point and flattened TC contacts are compared.

These measurements were conducted on a 6×6 in polycrystalline silicon wafer at the same furnace set points. It is noticeable that the flattened TC is sensitive enough to register signals related to aluminum melting at 660 °C and Al-Si eutectic freezing at 577 °C.

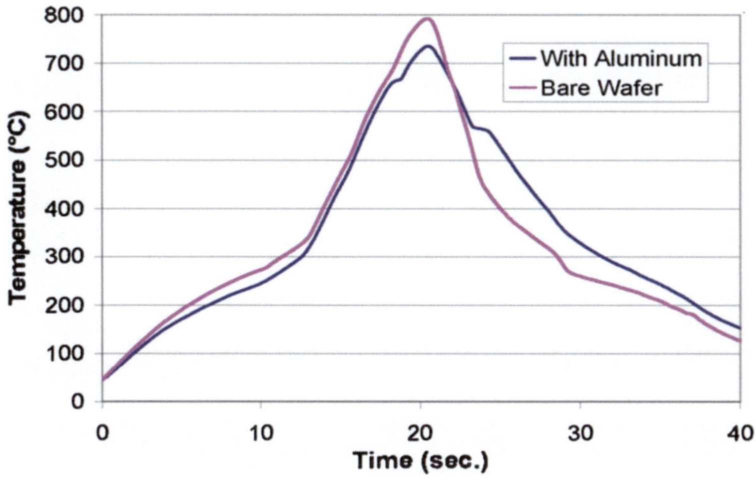

Fig. 6 Thermal profiles of bare and aluminum metalized wafers. The furnace set points were 400-400-500-700-800-880 at 200 ipm

After thorough data analysis, the TC junction design for the commercials units was finalized. In Fig. 5, a representative picture of the commercial unit is reproduced.

In Fig. 6, representative profiles for bare and aluminum metalized wafers recorded with a commercial unit are compared. The furnace set points for these measurements were the same as above. For these measurements also, 6×6 in polycrystalline wafers were used. Aluminum metalized wafer registered approximately 45 °C lower peak temperature.

Measurement Accuracy and Reproducibility

Two sets of data were collected to understand the measurement accuracy and repeatability. The first set of data was collected on bare and aluminum metalized wafers at ring condition p1, namely, 400-400-500-700-800-x at the belt speed of 200 ipm. To understand the influence of longer soak time, the second set of data was collected at firing condition p3, namely, 400-400-500-x-x-x at the belt speed of 200 ipm. The set points for the last zones (x) were varied to identify the minimum required value for M003 melting. For this measurement, the bare wafer weights of 10.5–10.6 gm and aluminum deposit weights of 1.5–1.6 gm were selected. For bare wafers, readings from five consecutive measurements were compared. On metalized wafers, six sets of data were collected. Data set 2–6 were used for the analysis.

In Fig. 7, TC2 readings of the two pro les recorded with a bare wafer are compared. On the day of measurement, the minimum required set point to commence M003 melting for p1 profile was 900 °C and for p3 profile was 870 °C.

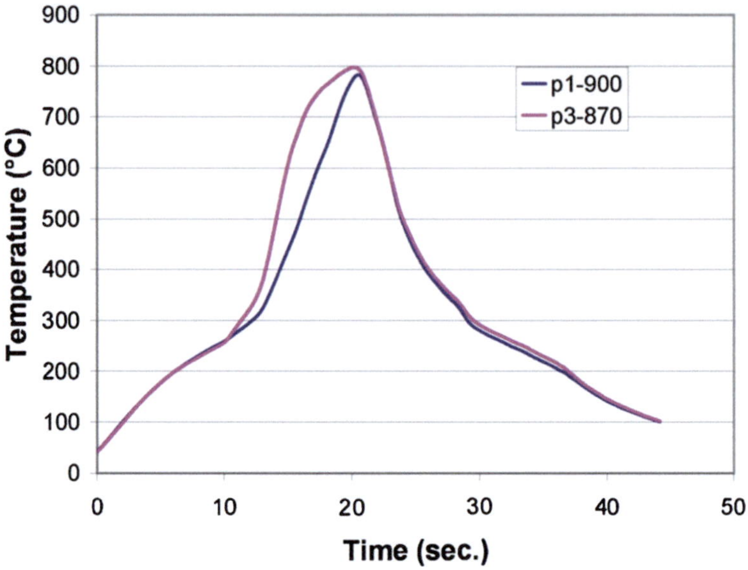

Fig. 7 Thermal profiles of bare silicon wafer at different set points. The set points for p1 are 400-400-500-500-700-800-900 at 200 ipm and for p3 are 400-400-500-870-870-870 at 200 ipm

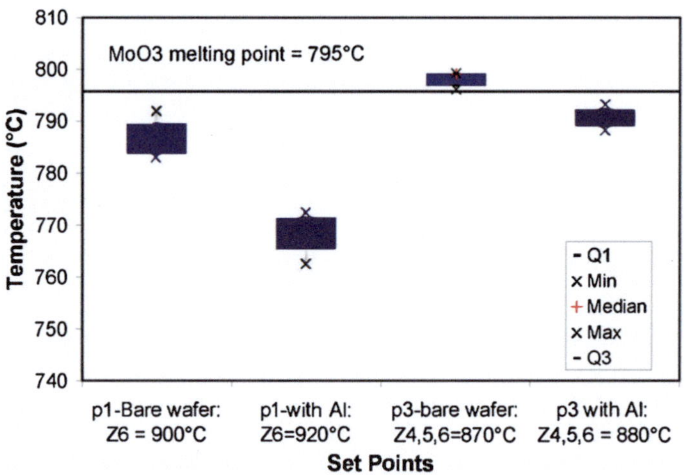

Fig. 8 Measurement repeatability

As seen, the thermocouples registered higher temperature when the soak time was longer.

In Fig. 8, the recorded peak temperatures of TC2 of all the measurements are compared. Both variables, namely, the presence of aluminum and the soak time at the peak temperature, influence the peak temperature readout.

Discussion and Highlights

Formation of front electrical contacts during the metallization heat treatment process involves several physical phenomena such as binder removal, glass melting, SiNx etching, Si etching, silver epitaxial film growth, etc. Some of these phenomena are sensitive to both temperature and the time at a specific temperature. In an IR furnace, all these phenomena occur at different parts of the heating and cooling segment but within a few seconds. To achieve the highest possible production yield, it is critical to maintain the entire profile as precisely as possible.

The actual temperature of the silicon wafer in an IR rapid thermal processing environment is influenced by several factors. The wafer temperature is a complex function of several factors such as IR absorption depth, thermal diffusivity, thermal mass, etc. IR absorption depth, in turn, is a function of factors such as silicon wafer crystallinity, doping level, and the IR radiation wavelength. Thermal mass of the wafer is primarily a fun ion of size and thickness of the wafer. In general, solar cell manufacturers metalize the wafers with various dimensions and doping levels.

The e-Clipse frame and SunKIC datalogger allow the users to test representative wafers from any manufacturing batch for the actual profile during the process to maintain quality and consistency.

Conclusions

In this paper, the salient properties of a user-friendly thermal profiling unit for silicon solar cells during the metallization process are presented. A calibration procedure for 800 °C processing is described. Based on the analysis, the following conclusions are drawn:

(a) Commonly used sheathed thermocouples record significantly lower peak temperatures than flattened TCs.
(b) Large surface area TC beads get closer to the correct peak temperature.
(c) At the same furnace set points, peak surface temperature of bare and aluminum metalized wafers differ, considerably.
(d) Flattened TCs read closer to the true peak temperatures as demonstrated by the melting of spots of M003.
(e) The KIC e-Clipse demonstrated higher accuracy and improved repeatability compared to conventional profiling methods.

References

1. Ballif, C., Huljic, D. M., Willeke, G., & Hessler-Wyser, A. (2003). *Applied Physics Letters, 82*(12), 1878–1880.

2. Khadilkar, C., Sridharan, S., Gnizak, D., Pham, T., Kim, S., & Shaikh, A. (2005) 2dh *EC PV solar energy conference*. Barcelona.
3. Neuhaus, D., & Munzer, A. (2007). *Advanced OptoElectronics*. Article ID 24521.
4. Borca-Tasciuc, T., Achimov, D. A., & Chen, G. (1998). *Materials Science Society Symposium, 525*, 103–108.
5. Vandenabeele, & Renken, W. (1998). *Materials Research Society Symposium Proceedings, 525*, 109–114.
6. Lee, J. I.. (2003). Rapid thermal processing of silicon solar cells. Ph. D thesis. Fraunhofer Institute of Solar Energy System.
7. Schubert, G., Huster, F., & Fath, P. *Proceedings of the 19th European photovoltaic solar energy conference (EU PVSEC'04)*, June 2004, p. 813.
8. Hilali, M. M., Rohatgi, A., & To, B. (2004). *Proceedings of the 14th workshop on crystalline silicon solar cells and modules*.

PID Temperature Control

Yuxiang Wang

Introduction

As you already learned in college courses, the temperature is a measurement of a material's internal molecular activity. As the molecules speed increases, the temperature of the material also increases. It is not only an important indicator in industrial areas but also closely related to our daily life. Therefore, to control temperature becomes necessary.

Generally, with measurement or detection of temperature change in space, we can adjust the amount of heat input or output to achieve an expected average temperature, and this whole process is called temperature control. In life, the air conditioner and heater might be the most familiar devices when we consider this term. In some industrial cases, precise temperature control by the reactor is critical for chemical reactions to take place. The function and purpose of these temperature control devices might be different, but they all follow the same design concept, which is to construct a control loop to reach thermostat. A simple thermostat system merely includes a heater and a cooler (air conditioner, etc.) and a switch to change mode between these two according to a thermometer. A more complex system adjusts the amount of heat added or removed, depending on the difference between the required temperature and the current temperature. This method is called proportional control. Furthermore, by using the accumulated signal (integral part) and error changing rate (derivative part), we can build a more complex PID controller which is mature in the industry. In this chapter, we will introduce different types of temperature control first and then talk about PID control in detail through its theory, methods, and modifications.

J. Xu et al., *Conveyor Belt Furnace Thermal Processing*,
https://doi.org/10.1007/978-3-319-69730-7_9

Different Types of Temperature Control

Control Loop Basics

Temperature control is essential in any case which a stable target temperature is required. There are two basic types of temperature control: open loop and closed loop control.

Open loop control is the most basic form, and the actual temperature does not influence the continuous heating/cooling at all. It is like the vehicle internal heating system. On a cold day, you may need to turn the heat on to full to warm the car to 25 °C. However, during a warmer time, the same setting would leave the inside of the car much warmer than the desired 25 °C (Fig. 1).

Closed loop control is far more complicated than the open loop, and the major difference between these two is the existence of *feedback.*

For example, closed loop control is analogous to a car with internal climate control. If you set the car temperature to 25 °C, the climate control will automatically adjust the heating (during cold days) or cooling (during warm days) as required to maintain the target temperature of 25 °C (Fig. 2).

In this example, the actual in-car temperature is the process variable (PV) , and the desired temperature 25 °C is the set point (SP). The difference between PV and SP is the error (*e*), which indicates whether the actual temperature is too high or too low and by how much.

The input to the process (the electric current in the heater) is the output from the controller. It is called the manipulated variable (MV) or the control variable (CV). To sum up, by subtracting the sensed actual temperature (PV) and the target temperature (SP), we get the error (*e*), and according to it, the controller calculated how much current to supply to the heater (MV).

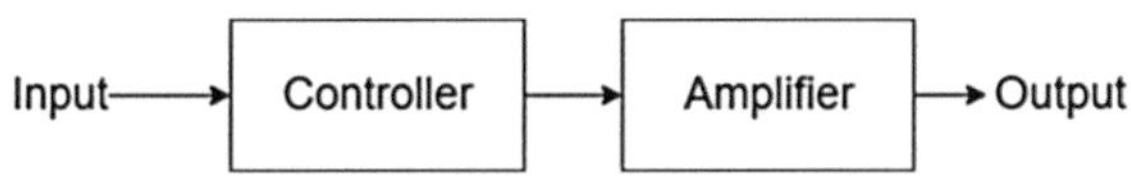

Fig. 1 Open loop block diagram

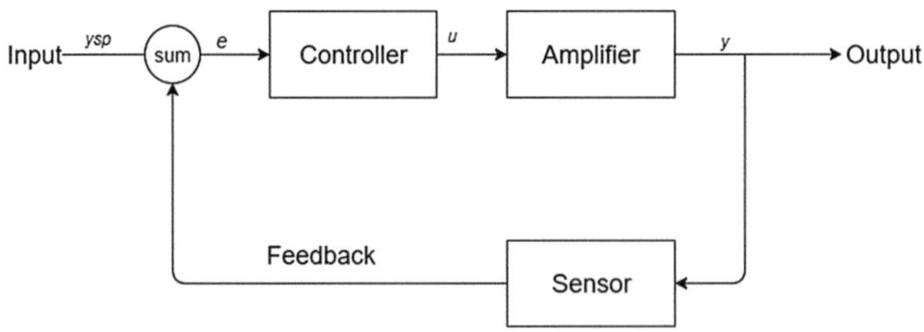

Fig. 2 Closed loop block diagram

To simplify, let us assume the process variable increases when the manipulated variable increase. The principle of feedback can be expressed as follow:

Increase MV when PV is smaller than SP, and decrease MV when PV is larger than SP.

This kind of feedback is called negative feedback as the manipulated variable moves in opposite way to the process variable.

Definition

Process variable is the current status of a process under control.

Set point is the desired or target value for an essential variable of a system.

On-Off Control

The on-off control is the simplest form of a control algorithm. The output from the device is either on or off, with no intermediate state. Its mechanism can be described mathematically as below:

$$u = \begin{cases} u_{\max} & \text{if } e > 0 \\ u_{\min} & \text{if } e < 0 \end{cases} \tag{1}$$

where e is the control error between set point temperature and the actual temperature. The characteristics of on-off controller determine that maximum corrective action is always applied. Therefore, the manipulated variable u is highest when e is positive and lowest when e is negative. An on-off controller often fulfills the needs of keeping the actual temperature close to the set point. However, it will significantly result in a system with oscillating variables. If we look back to Eq. 1, the control variable is not defined when the error is zero. In a case where the cycling is frequent, it is necessary to do some modification to prevent damage to contactors and valves, by either hysteresis or a dead zone (Fig. 3).

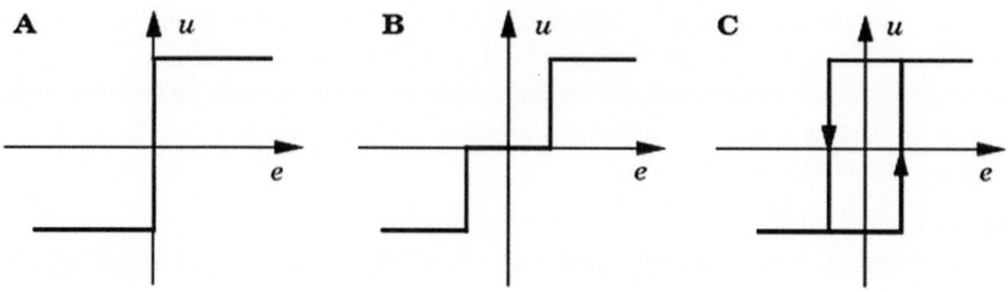

Fig. 3 Controller characteristics for ideal on-off control (**a**), modifications with dead zone (**b**) and hysteresis (**c**)

Proportional Control

On-off control often results in the system overreaction, because the small error change will make the manipulated variable change over the full range. In proportional control, we can prevent this influence by making the feedback proportional to the control error.

$$u = K\left(ysp - y\right) = Ke \tag{2}$$

where K is the controller gain.

When using temperature controller, when the temperature approaches the set point, the proportional controller will decrease the average power supplied to the heater. This slows down the heater so that the temperature will not overshoot but reach the set point gradually. By changing the time proportion of on-time and off-time, we can control the temperature more smoothly. The proportional effect occurs within the "proportional band" around the set point temperature. Outside this band, the controller is used as a switching unit; the output is fully open (below the frequency band) or completely closed (above the frequency band). However, in the frequency band, the output is turned on and off at a ratio of the measured difference to the set value. At the set point, which is also the midpoint of the proportional band, the output on-off ratio is 1: 1. Therefore, the on-time and off-time are equal here. If the temperature is much higher than the set value, the switching time is proportional to the temperature difference.

PID Control

The third type of controller is a combination of proportional, integral, and differential control (PID). The controller cooperates proportional control and two additional adjustments, which helps the machine to compensate for changes in the system automatically. These adjustments, integrals, and derivatives are expressed in time domain. Proportional, integral, and derived terms must be tuned individually to the specific system using trial and error. It is the most accurate and stable controller of these three and is best suited for systems with relatively small masses that react quickly to changes in process variable (energy, volume, etc.) added to the process. More detail will be introduced in next section.

Definition

The proportional term produces an output value that is proportional to the current error value.

The integral term is the sum of the instantaneous error over time and gives the accumulated offset that should have been corrected previously.

The derivative term of the process error is calculated by determining the slope of the error over time and multiplying this rate of change by the derivative gain K_d.

PID Control

Introduction

The PID controller is the most commonly used control algorithm so far. Its history can be traced back to nineteenth century on a speed governor design. Nowadays, most feedback loops use this algorithm or its variations. It is implemented in many different forms, as a controller itself or as a part of furnace temperature control or a neutralization pH control system. This chapter gives an introduction to PID control, including the basic algorithm and several representations.

PID Controller Theory

The PID control combines proportional control with integral and derivative terms. The sum of these terms above constitutes the manipulated variable (MV). $u(t)$ is defined as the output of PID controller here, and the ideal form of this algorithm is

$$u(t) = \text{MV}(t) = K_\text{p} e(t) + K_i \int_0^t e(\tau) d\tau + K_\text{d} \frac{de(t)}{dt} \tag{3}$$

where

- K_p is the proportional gain, a tuning parameter.
- K_i is the integral gain, a tuning parameter.
- K_d is the derivative gain, a tuning parameter.
- $e(t)$ = SP – PV(t) is the error (SP is the set point, and PV(t) is the process variable).
- t is the time or instantaneous time (the present).
- T is the variable of integration (takes on values from time 0 to the present t).

Slightly different from the ideal form, and most often encountered in industry, the standard form is

$$\text{MV}(t) = K_\text{p} \left(e(t) + \frac{1}{T_\text{i}} \int_0^t e(\tau) d\tau + T_\text{d} \frac{d}{dt} e(t) \right) \tag{4}$$

where

- T_i is the integral time.
- T_d is the derivative time.

In this standard form, the parameters have a clear physical meaning. The inner summation produces a new single error value which is compensated for future and past errors.

Equivalently, the transfer function in the Laplace domain of the PID controller is

$$L(s) = K_p + K_i / s + K_d s, \tag{5}$$

where s is the complex frequency.

Proportional Term

The proportional term gives an output value that is proportional to the current error value. By multiplying the error by a constant K_p, the proportional gain constant, we can adjust the proportional response. The proportional term is given by

$$P_{out} = K_p e(t). \tag{6}$$

A high K_p value results in a significant change in the output for a given change in the error. Which means, the higher the proportional gain, the more unstable the system can be. In contrast, a small K_p value results in a small output response to a significant input error and a less sensitive controller and vice versa. Tuning theory and industrial practice indicate that the proportional term should contribute the bulk of the output change.

Steady-State Error

When the system approaches the set point, the error gets smaller and smaller. As Eq. 4 shows above, the proportional term is the product of error and K_p; the error will infinitely decrease but never reach zero. Therefore, a non-zero error is required to drive it. Steady-state error (SSE) is directly proportional to the process gain and inversely proportional to K_p. SSE may be mitigated by adding a compensating bias term to the set point or output or corrected dynamically by adding an integral term.

Integral Term

Mathematically, the integral term is the product of integral gain K_i and the accumulated error in given time. Therefore, the integral term is directly proportional to both the magnitude of $e(t)$ and time duration. The sum of the instantaneous error in the time duration is corrected by K_i and added to the controller output.

The integral term is

$$I_{\text{out}} = K_{\text{i}} \int_{0}^{t} e(\tau)\,d\tau \tag{7}$$

The integral term helps to speed up the process toward set point and eliminates the remaining SSE that occurs with a simple proportional controller. One thing to note, because the integral term is a form of past accumulated errors, it may not represent the current state perfectly, which causes the present value to overshoot the set point value.

Derivative Term

The purpose of the derivative term is to improve the closed-loop stability. The instability mechanism can be described intuitively as follows. Because of the process dynamics, it will take some time before a change in the control variable is noticeable in the process output. Thus, the control system will be late in correcting for an error. The action of a controller with the proportional and derivative action may be interpreted as if the control is made proportional to the predicted process output, where the prediction is made by extrapolating the error by the tangent to the error curve.

The derivative of the process error is calculated by determining the slope of the error over time and multiplying this rate of change by the derivative gain K_d. The magnitude of the contribution of the derivative term to the overall control action is termed the derivative gain, K_d [6].

The derivative term is given by

$$D_{\text{out}} = K_{\text{d}} \frac{de(t)}{dt} \tag{8}$$

The derivative action predicts the behavior of the system, thus improving the setting time and stability of the system. The ideal derivative is not causal, so the implementation of the PID controller includes additional low-pass filtering of the differential term to limit the high-frequency gain and noise. Derivatives are rarely used in practice due to their different effects on system stability in real-world applications.

Loop Tuning

Tuning the control loop is a process of pursuing optimum value by adjusting its control parameters (K_p, K_i, K_d) to the desired control response. Stability is a basic requirement; however, since requirements differ from case to case, requirements may conflict with each other.

PID tuning is a challenge, even if there are only three parameters, because it must meet the complex PID control constraints. Therefore, there are a variety of loop adjustment methods, and more sophisticated techniques are the subject of patents; this section describes some of the traditional manual methods for loop tuning.

Designing and tuning the PID controller seem conceptually intuitive, but it may be difficult to achieve multiple objectives, such as short transient and high stability. The PID controller typically uses the default tuner to provide acceptable control, but the performance can be improved by careful adjustment, and poor tuning could lead to unacceptable performance. The initial design usually needs to be repeatedly tuned by computer simulation until the closed-loop system executes or compromises as expected.

Some processes have a certain degree of nonlinearity, so the parameters that work well under no-load conditions do not function at the no-load start; this can be corrected by using different settings in different operating areas.

Stability

If the parameters of the PID controller are incorrectly selected, the controlled process input might be unstable (i.e., its output diverges, oscillates or not, and is limited only by saturation or mechanical damage). Instability is caused by excess returns, especially in the presence of significant hysteresis.

In general, we need a stable response, and for any combination of process conditions and set points, the process must not oscillate, although sometimes the marginal stability (bounded oscillation) is acceptable or desirable.

In the Laplace domain, it is easier to see the origin of instability.

The total loop transfer function is

$$H(s) = \frac{K(s)G(s)}{1 + K(s)G(s)} \tag{9}$$

where

- $K(s)$, PID transfer function
- $G(s)$, plant transfer function

The system becomes unstable where the closed-loop transfer function diverges for some values of s. This situation occurs in the case of $K(s)G(s) = -1$. Normally, this happens when a 180° phase shift occurs. For the frequency of high phase shift, stability is ensured when $K(s)G(s) < 1$. This more general formalism of this effect is called the Nyquist stability criterion.

Optimum Behavior

The variation of applications determines the optimum behavior of process change or set point change.

The two primary requirements are regulation (interference suppression—stay at a given set point) and command tracking (to achieve set point changes); these indicate the degree to which the control variable tracks the expected value. The specific criteria for command tracking include rise time and settling time. Some processes cannot allow overdosed process variables that exceed the set value, for example, if this will be unsafe. Other processes must minimize the amount of energy consumed by the new set point.

Overview of Methods

There are several ways to adjust the PID loop. The most efficient method usually involves the development of some form of process model and then selects P, I, and D according to the dynamic model parameters. Manual tuning methods can be comparatively time-consuming, especially for systems with longer loop time.

The choice of method significantly depends on the system response time and whether the loop can be "off-line" for tuning. If an "off-line" loop is possible, the optimum tuning method typically involves making the system a step change for input, measuring the output as a function of time, and using the response to determine the control parameters.

Manual Tuning

If the system must remain online, one way is to first set the K_i and K_d values to zero. Increase K_p until the output of the loop oscillates, and then in the "quarter-amplitude attenuation" type response, K_p should be set to about half of the value. Then increase K_i until any offset is corrected for a sufficient period. However, too many K_i will lead to instability. Finally, if necessary, increase K_d until the loop can quickly reach its reference value after the load disturbance. However, too much K_d may result in excessive response and overshoot. The quick PID loop adjustment is usually slightly overshot to achieve the set value faster; however, some systems cannot accept overshoot, in which case an overdamped closed loop system is necessary, which will

Table 1 Effects of increasing a parameter independently [1]

Parameter	Rise time	Overshoot	Setting time	Steady-state error	Stability
K_p	Decrease	Increase	Small change	Decrease	Degrade
K_i	Decrease	Decrease	Increase	Eliminate	Degrade
K_d	Minor change	Decrease	Decrease	No effect in theory	Improve if K_d is small

need a K_p setting that is significantly smaller than causing half of the K_p setting for oscillation (Table 1).

Ziegler-Nichols Method

Developed by John G. Ziegler and Nathaniel B. Nichols 80 years ago, the Ziegler-Nichols tuning method is a heuristic method for adjusting the PID controller. Firstly, it sets the I (integral), and D (differential) gains to zero. Then, make the "P" (proportional) gain K_p increase from zero until *the ultimate gain* K_u is approached, and the output of the control loop has steady and consistent oscillations. K_u and the oscillation period T_u are used to set the P, I, and D gains according to the type of controller used (Table 2).

These three parameters are used to establish the correction $u(t)$ from the error $e(t)$ via the equation:

$$u(t) = K_p \left(e(t) + \frac{1}{T_i} \int_0^t e(\tau)\, d\tau + T_d \frac{de(t)}{dt} \right) \tag{10}$$

where T_i is the integral time and T_d is the derivative time.

PID controller in the industry often encountered in the standard form. In this form, the K_p gain is applied to the Iout and Dout terms.

Ziegler-Nichols tunes created a "quarter-wave decay." This method is acceptable in some cases, but not the best for all applications.

The tuning rules are designed to give the PID loop the best anti-jamming capability.

Also, it produces positive gain and overshoot, which brings problems when some applications want to reduce or eliminate overshoots.

Table 2 Ziegler-Nichols method parameter settings [2]

Control type	K_p	T_i	T_d
P	$0.5K_u$	–	–
PI	$0.45K_u$	$T_u/1.2$	–
PD	$0.8K_u$	–	$T_u/8$
Classic PID [2]	$0.6K_u$	$T_u/2$	$T_u/8$
Pessen integral rule [2]	$0.7K_u$	$T_u/2.5$	$3T_u/20$
Some overshoot [2]	$0.33K_u$	$T_u/2$	$T_u/3$
No overshoot [2]	$0.2K_u$	$T_u/2$	$T_u/3$

PID Tuning Software

Most modern industrial facilities no longer use the above manual calculation method to tune the loop. Pursuing more consistent results, engineers and scientists tend to use PID tuning and loop optimization software instead. These packages will collect data, develop process models, and propose optimal tuning. Some software packages can even be developed by collecting data from reference changes.

Mathematical PID loop tuning initiates a pulse in the system and then uses the frequency response of the controlled system to design the PID loop value. In loops with a response time of several minutes, it is advisable to use mathematical loop tuning because the tests and errors may take several days to find a stable set of loop values. It is never easy to find the optimum value. Some digital loop controllers provide self-tuning functions, which send tiny set point changes to the process, allowing the controller itself to calculate the best tuning value.

Other formulas can be used to adjust loops according to different performance criteria. Many patent formulas are now embedded in PID tuning software and hardware modules.

The progress of the automatic PID loop tuning software also provides an algorithm for adjusting the PID loop in a dynamic or unsteady (NSS) scenario. The software will model the dynamics of the process by interfering with the process and calculate the PID control parameters (Fig. 4).

Limitations of PID Control

While PID controllers are suitable for many control problems, and are often performed satisfactorily without any improvement or only coarse adjustment, they sometimes perform poorly in some applications and do not provide optimal control. The fundamental difficulty of PID control is that it is a feedback control system with constant parameters, without a direct understanding of the process, so the overall performance is a simulation based on reaction and comptonization. Although PID control is the best controller for observers without process models, better

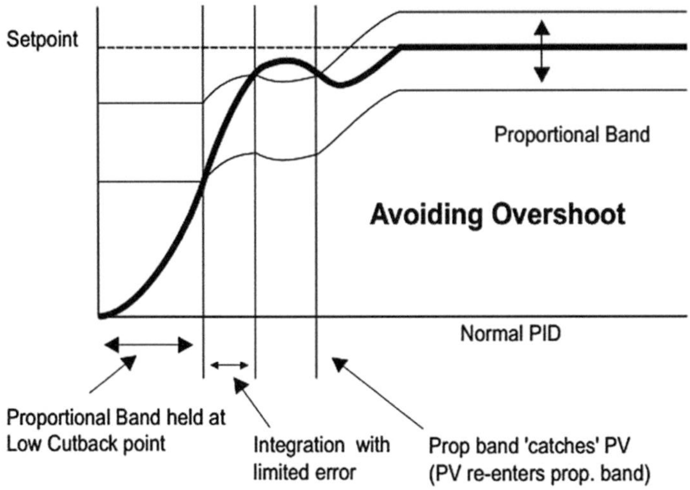

Fig. 4 Auto software tuning

performance can be achieved by explicitly modeling the actors of the process without relying on a monitor.

When the PID controller is used alone, the PID controller must reduce the PID loop gain to prevent the control system from overshooting, oscillating, or tracking the control set point value. The presence of nonlinear problems is another form of difficulty, which may trade off regulation versus response time, barely react to frequent process behavior changes (e.g., process changes after warming up), and delay in responding to great disturbances.

The most notable improvement is the combination of feedforward control and system knowledge and only use the PID to control the error. Alternatively, the PID can be modified in a smaller way, such as by changing the parameters (gain adjustment in different use cases or adaptive modification based on performance), improved measurement (higher sampling rate, accuracy and precision, and low-pass filter if needed), or cascade multiple PID controllers.

Linearity

Another problem faced by PID controllers is that they are linear, especially symmetrical. Therefore, the performance of the PID controller in a nonlinear system is variable. For example, in temperature control, the usual design only has heating elements but not cooling, so overshoot can only be slowly corrected—cannot be forced down. In this case, the PID should be adjusted to overdamped to prevent or reduce overshoot, although this performs worse (increases the settling time).

Noise in Derivative

Another problem is about the derivative term. The derivative term amplifies higher frequency measurements or process noise, which may cause a significant number of changes in output. Filtering the measurement by a low-pass filter is usually helpful to remove high-frequency noise components. Since low-pass filtering and differential control can cancel each other, the amount of filtering is limited. So, the low noise instrument can be substantial. Using a nonlinear median filter can improve filter efficiency and actual performance. Sometimes, the differential band can be turned off, and almost no control is lost. In this case, the PID controller is used as a PI controller.

Modification of PID Control

Integral Windup

Integral windup refers to the case where a large set point change (e.g., positive change) occurs in the PID feedback controller. The integral term accumulates on the rise, thus overshooting and continuing to increase because the cumulative error is unwound (offset by errors in another direction). The problem is the excessive overshoot.

This issue can be solved by:

- Initializing the controller integral to the desired value, for example, the value before the problem
- Increasing the set point in the appropriate ramp
- Disabling the integral function until the controlled process variable (PV) enters the controllable area
- Preventing integral term from accumulation above or below the intended range for integral term
- Back-calculating the integral term to limit the process output within the available range

Overshooting from Known Disturbances

For example, the PID loop is used to control the temperature of the electric resistance furnace with a stabilized system. Now when the door is opened, and the coolant is placed into the furnace, the temperature drops below the set value. The integral feature of the controller tends to compromise for errors by introducing another error in the positive direction. This overshoot can be avoided by temporarily turning off the integral function after the door has been opened when the control loop requires a reheating furnace.

Future Pattern of PID Control

From the theory of automatic control to the modern control theory, we have completed the transformation from simple system control to the complex control system (input and output variables increase, control accuracy increases). From modern control theory to intelligent control, we have completed the transition from precision research to controlling specific functions (we increasingly want to control humanization rather than accuracy). Nowadays, the computer science is popular, especially in machine learning, deep learning, and big data; the development of cloud computing to control has brought new ideas. Considering the new trends, we can say the future direction of PID control will tend to cooperate with learning system. Therefore, the next trend may very well focus on the pursuit of performance instead of precision. Control in many cases is not only a matter of precision. Take control of the washing machine as an example. When we add different clothes, our purpose is not the same washing time, but different time according to the number of clothes and how dirty they are. It is hard to say the best results for such problems. The intelligent control method is to put previous experience data to the controller so that the controller can determine the best washing way. Therefore, the future focus may be closer to the data and continually optimize the algorithm to achieve a reasonable performance (rather than precision performance).

References

1. Ang, K. H., Chong, G. C. Y., & Li, Y. (2005). PID control system analysis, design, and technology (PDF). *IEEE Transactions on Control Systems Technology, 13*(4), 559–576.
2. Ziegler, J. G., & Nichols, N. B. (1942). Optimum settings for automatic controllers (PDF). *Transactions of the ASME, 64*, 759–768.
3. Åström, K. J., & Hägglund, T. (1995). *PID controllers: Theory, design, and tuning*. Research Triangle Park: Instrument Society of America.
4. Qadir, A. F. M. S. (2013). Electro-mechanical modeling of SEDM (separately excited DC motor) & performance improvement using different industrial controllers. https://www.amazon.com/Electro-Mechanical-Separately-Performance-Improvement-Controllers/dp/1304227650
5. Cooper, D. Integral (reset) windup, jacketing logic and the velocity PI form. http://controlguru.com/integral-reset-windup-jacketing-logic-and-the-velocity-pi-form/
6. Zehntner, U. (2004). Encyclopedia of dairy sciences. *International Dairy Journal, 14*, 269. https://doi.org/10.1016/j.idairyj.2003.12.001.

Introduction to Materials and Firing Parameters in Thick Film Firing

Lyman Li

Introduction

Thick film technology is a type of technology which uses conductive, resistive, and insulating pastes, deposited in patterns defined by screen printing and fused at high temperature (500–1000 °C) onto a ceramic substrate permanently. The thickness typically ranges from 5 to 20 μm.

The substrates are passed through a continuous belt furnace with multiple firing zones to be fired. Inside the tunnel, the parts are going through different phases including organic burnout, softening/melt, resistive compound formation, solidification, and, at last, annealing. The substrates are able to absorb energy from infrared sources because infrared furnaces can reach to high temperature rapidly and control temperatures precisely. By using an infrared furnace, the fire rates can be increased by three to four times than using a conventional convection furnace.

Material

The most commonly used substrate materials are alumina ceramic with 94–98% alumina content. The particle sizes range from 3 to 5 μm. Unlike thin film technology, smooth surface finish is not required for thick film technology, and "as-fired" ceramics are suitable. Polished ceramics or glass is acceptable. The insulation resistance of the glass is very important, including its behavior at high temperature (500–1000 °C). Low percentages of low free alkali are required. More and more materials are being used as substrates, such as porcelainized steel, organic materials (epoxies, flexible substrates), and even synthetic diamond.

Inks or pastes are added sequentially to produce required conductor patterns and resistor values on the substrate above. With different formulas, the pastes can be

J. Xu et al., *Conveyor Belt Furnace Thermal Processing*,
https://doi.org/10.1007/978-3-319-69730-7_10

used to produce conductors, resistors, and dielectrics which contain binders, carriers, and materials to be deposited (typically pure metals, alloys, and metal oxides).

After firing, the metal particles are bound together to the substrate by the glassy phase, and this is particularly important at the substrate-ink interface. On a microscale, fired surfaces are usually not even or homogeneous. This can make the next step—wire bonding—very difficult. The particle sizes are smaller than solder pastes, so the suspensions are more stable. The inks must be at appropriate viscosity for the screen printing process. It must fall between solid and liquid. If the viscosity is too low, the inks will spread after printing; if the viscosity is too high, there will be mesh marks which is a failure of leveling.

No matter what printing method is used, the obvious lower limits on feature sizes will make high-density designs difficult. At the same time, poor edge definition of narrow tracks will have impacts on high-frequency performance. So some materials are developed for printing over the whole substrate area and subsequent etching.

Conductor Pastes

In hybrid circuit manufacturing, conductor pastes are the largest proportion of pastes that are used. As materials which are used to transfer signals between circuits, resistor terminations, crossover connections, discrete component attach points, wire bond attach points, and so on, conductor pastes must have low resistivity. Some other requirements are good solderability, compatibility with other components, good adhesion to the substrate, good wire bondability, cost/performance trade-off, and so on. A thick film conductor contains three functional materials: metal powders, permanent binders, and the vehicle. As for metal powders, noble metal materials include gold, silver, palladium, and platinum and/or alloys of them, while copper is the only well-fabricated material to be used in products at this time.

Resistor Pastes

Modern resistor pastes are grounded on oxides of ruthenium, iridium, and rhenium. Compared to the earlier pastes, these materials are stable during the firing process. They also provide better temperature coefficient of resistance.

Dielectrics Pastes

The lowest possible dielectric constant is to be preferred in order to minimize the capacitance associated with crossovers. A high dielectric constant is preferred for the fabrication of capacitors. A good paste will require high dielectric strength, good insulation resistance, and low power factor.

Paste Selection

It is very important to select a proper paste for circuits so that desirable performance could be achieved. Noble metal pastes seem to be the primary choice at this moment for reliable and excellent results. In the research area, the medium temperature base metal pastes are more popular. However, the low-cost polymer with lower performance is still suitable in many commercial products. Some factors listed below need to be considered when selecting a paste.

Conductivity: From low to high sheet resistivity, there are copper, silver-platinum, gold, silver-palladium, and gold-palladium.

Firing temperature: The maximum capacity of most furnaces is less than 1000 °C. Hengli HSK furnace can heat up to 1050 °C.

Adhesion: It is the strength required to pull the conductor from the substrate. The solder joint/conductor has to carry the weight of any added chip components.

Interconnection: Wire bonding and soldering are commonly used to connect fired circuits with other components.

Firing Parameters

After printing the paste on the ceramic substrates with a given circuit pattern, the substrates are leveled at room temperature of about 10 min and then dried at low temperature (150 °C) for 15 min in order to remove the organic solvents by evaporation. The drying process can be achieved by passing printed ceramics through the belt furnace, so the printed ink is frozen and ready to be fired. Many firing parameters and conditions such as heating rate, belt speed, atmosphere, and material types will influence circuit properties such as film thickness, sheet resistivity, solder leach resistance, adhesion, and so on. When tuning the thick film furnace parameters, three major factors are very important to controlling the temperature profile: temperature, airflow, and firing time in each zone. In his study on the manufacturing process of Ag/Pd thick film conductor circuits, Yen-Chang Tseng analyzed the influence of the three factors on sheet resistivity.

60 Minute Profile			
Control Point	Nominal	Tolerance	Range
100C - 100C	55 min	+/- 5 min	50 - 60 min
600C - 600C	31 min	+/- 2 min	29 - 33 min
800C - 800C	19 min	+/- 1 min	18 - 20 min
850C Dwell (Time)	10 min	+/- 0.5 min	9.5 - 10.5 min
850C Dwell (Temp)	850C	+/- 3C	847C - 853C
300C - 500C Ascent	45C/min	+/- 10C/min	35C - 55C/min
800C - 600C Descent	30C/min	+/- 5 min	27C - 33C/min
700C - 300C Descent	50C/min	+/- 10C/min	40C - 60C/min
Burnout Removal	500C	+/- 50C	450C - 550C

30 Minute Profile			
Control Point	Nominal	Tolerance	Range
100C - 100C	33 min	+/- 3 min	30 - 36 min
600C - 600C	19 min	+/- 1 min	18 - 20 min
800C - 800C	14 min	+/- 1 min	13 - 15 min
850C Dwell (Time)	10 min	+/- 0.5 min	9.5 - 10.5 min
850C Dwell (Temp)	850C	+/- 3C	847C - 853C
300C - 500C Ascent	95C/min	+/- 10C/min	85C - 105C/min
800C - 600C Descent	70C/min	+/- 10C/min	607C - 80C/min
700C - 300C Descent	95C/min	+/- 10C/min	80C - 105C/min
Burnout Removal	500C	+/- 50C	450C - 550C

Fig. 1 Recommended thick-film firing profiles by DuPont

Temperature Profile

Temperature profile is the most significant factor which affects response. High-temperature firing will produce very low resistivity conductor circuits. Medium-temperature firing can also give ideal results. Considering the firing process is steadier at medium temperature, medium-temperature firing is recommended. If minimum resistivity is needed, high temperature can be used.

Figure 1 is the recommended profiles by DuPont.

Airflow

The leading thick film material manufacturer DuPont emphasized that proper furnace setup is critical for successful high-yield thick film processing. Per DuPont, the two key parameters that control successful firing are furnace air and the firing profile.

DuPont believes that the airflow is important due to the following six reasons:

(a) A sufficient quantity of clean, dry air must be present for thick film to fire correctly.
(b) Intake air should be drawn from a point 6–10 ft above the roof line and as far away as possible from contaminant-containing exhausts.
(c) Oil-free compressors should be used whenever possible.
(d) A series of dryers and filters must be used to dry to a dew point of −30 °C at 80–100 psi. Standard cooling dryers in conjunction with heated desiccant dryers will achieve this value.
(e) Activated charcoal filters will help to remove halogen and sulfur-type contaminants.
(f) Micron and submicron filters will remove particulate matter.

DuPont further recommends a formula to *determine how much air is required in the furnace:*

V = PLAWS where:

- V = The volume of air in liters/minute required in the burnout section of the furnace
- P = Printed area of the most dense circuit expressed as a decimal
- L = Belt load factor. How much of the burnout section is full when parts are loaded, expressed as a decimal
- A = Constant of 0.4 l/cm^2
- W = Belt width (cm)
- S = Belt speed (cm/min)

The amount of air required in the firing section is 10–20% higher than for the burnout section.

To set up the airflow properly:

- Turn off all flowmeters.
- Turn up the burnout air to the value calculated above.
- Turn the firing air to its value (1.1–1.2 times PLAWS).
- Check to make sure air is flowing out the entrance and exits of the furnace.
- Turn up the Venturi exhaust until room air is just being drawn into the furnace; this will ensure that all air introduced to the furnace is removed.
- Turn up the entrance and exit air curtains until flow at the exit and entrance is out of the furnace. This will prevent air from being drawn into the furnace.

Overall, DuPont's recommendations work well. Sometimes some customers complained about too high usage of dry compressed.

To minimize the usage, one can develop a plan to monitor the fired product quality. Then the airflow into the furnace can be reduced gradually and still maintains the positive pressure inside the furnace, like a clean room design, to keep the furnace inside air clean. In most cases, the airflow to the entrance and exit curtains can be reduced too. The Venturi-assisted exhausts can be kept at minimum, as long as the furnace chamber has positive pressure.

References

1. Thick film processing. http://www.furnacepros.com/Furnaces/thickfilm.php
2. Thick film technology. http://www.idc-online.com/technical_references/pdfs/electronic_engineering/Thick_film_technology.pdf
3. Owner's manual of HSK2505–0611, Hengli thick film firing furnace.
4. Yen-Chang, T. (1995). Manufacturing process analysis of Ag/Pd thick film conductor circuits, Lehigh Preserve.
5. Taylor, B. E., Felten, J. J., & Larry, J. R. (1980). Progress in technology of Ag thick film conductors. *IEEE Trans, CHMT-3*(4), 504–517.
6. Naguib, H. M., & MacLaurin, B. K. (1979). Ag migration and the reliability of Pd/Ag conductors. *IEEE Trans, CHMT-2*(2), 196–199.
7. Roffey, M., & Metke, N. (1990). Atmospheres and their effects on air firing thick films. *Hybrid Circuit Technology-7*(11), 45–50.
8. Proper furnace setup in Dupont website. http://www.dupont.com/products-and-services/electronic-electrical-materials/articles/proper-furnace-setup.html

Influence of Belt Furnace on Engine Valve Heat Treatment

Sean Song

Introduction

Internal combustion engines contain multiple cylinders; each cylinder has one intake valve and one exhaust valve. Both valves open and close internal passages in the cylinder head. The intake valve is the larger of the two valves. It controls the flow of fuel into the combustion chamber. The exhaust valve controls the flow of exhaust gases out of the cylinder.

Engine valves are essential parts for engine functioning. They are located in the cylinder head and can be classified into intake valves and exhaust valves. The intake valves bring in air/fuel into the chamber for combustion, and the exhaust valves let exhaust out after burning.

Figure 1 shows typical engine valves.

The open and close of valves are decided by the cylinder piston positions. A detailed explanation by Marshall Brain on how engine works can be found at http://auto.howstuffworks.com/engine4.htm.

Engine Valve Heat Treatment Description

The working conditions of engine valves are severe. Doug Kaufman states that intake valves typically run at the range of 800–1000 °F(427–538 °C), while exhaust valves typically run at 1200–1450 °F(649–788 °C), due to the temperature difference of intake gas and exhaust. Besides high temperature, valve also experiences cyclic loading. A valve can open and close dozens of times per second.

Because of the high working temperature and strength required in such conditions, heat-resistant steels are often used. Besides proper material selection, proper heat treatment is essential in manufacturing high performance engine valves. It is

J. Xu et al., *Conveyor Belt Furnace Thermal Processing*,
https://doi.org/10.1007/978-3-319-69730-7_11

Fig. 1 Typical engine valves (Courtesy of online picture http://www.plxsport.com/atv-parts-valvetrain-wiseco-titanium-intake-valve-suzuki-ltr-450?att_id=0)

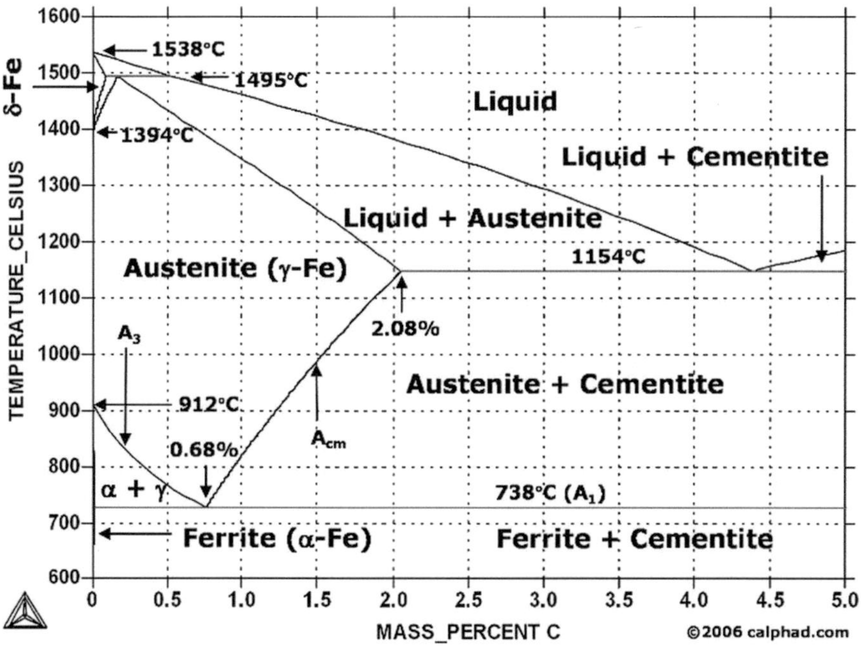

Fig. 2. Fe-C phase diagram (Courtesy of online source http://www.calphad.com/iron-carbon.html)

vital to create the desired properties like strength, wear resistance, toughness, fatigue strength, hardness, and microstructure.

Normal heat treating methods include annealing, normalizing, tempering, and hardening. To decide a specific heat treatment process, alloy phase diagram (Fig. 2) is the fundamental tool.

A typical heat treating process for engine valves of VAZ passenger cars was published in *Metallovedenie i Termicheskaya Obrabotka Metallov*, No. 10, pp. 6–9, 1996. Several major steps in the heat treatment process for inlet valves include (1) preliminary heat treatment, annealing to around 700 °C for 3–4 h to obtain spheroidized pearlite with certain hardness and (2) stabilizing annealing. Hold 2–3 h at 600–620 °C to release stress and eliminate warping issues that might occur in other process.

For outlet valves the heat treatment process is a little different because of the quality requirement difference between inlet and outlet valves.

For this application, a modified nitrogen capable fast firing furnace is used.

A fast fire furnace heats from ambient to 1050 °C in approximately 40 min and is designed to sustain continuous on/off heating and cooling cycles resulting from alternating periods of production and nonuse. It features an ultraclean low-mass refractory heating chamber equipped with FEC (fully enclosed coil) heaters formed into ceramic insulation panels.

With the use of advanced insulation materials, lower thermal capacity enables the furnace to warm up and cool down very quickly and lose less heat to the environment.

The belt would travel through the furnace on a metal hearth plate with side walls, which would prevent products from falling off the belt.

References

1. Funatani, K. (2004). Heat treatment of automotive components: Current status and future trends. *Transactions of the Indian Institute of Metals, 57*(4), 381–396.
2. Doug Kaufman. Understanding valve design and alloys.
3. IIT Bombay. Design for heat treatment, online presentation.
4. Cherdantsev, A. N., Makar'ev, A. N., Akhant'ev, V. P., & Kaplina, I. N. (1996). Technology for heat treatment of engine valves of VAZ passenger cars. *Metallovedenie i Termicheskaya Obrabotka Metallov, 10*, 6–9.

The Influence of a Belt Furnace on the Brazing Process

Introduction

"Brazing" refers to a process that joins metals together. By using a braze alloy or filler metal to unite individual pieces, a strong metallurgical bond is created without melting or changing the essential structure of the original materials. With proper technique, the connection between the metals is often stronger than the actual metals themselves, which ensures the longevity of the bond. While brazing is one of the oldest processes for joining metals, technological progress has opened the door for brazing other materials such as ceramic or glass together. Brazing is considered to be essential to many industries, including automotive, aircraft, aerospace, electrical, and more. Additionally, brazing serves four major purposes for electronics: mechanical, electrical, hermetic, and thermal. Although there are different techniques for brazing, furnace brazing specifically allows for high rates of commercial-scale production.

Furnace brazing has many specific advantages. The metals being joined do not have to be the same; dissimilar metals with varying thicknesses can also be united. This allows for complex arrangements of different parts, without distorting the original metals. Multiple joints can also be brazed at the same time. Moreover, superior products can be achieved by brazing in a furnace, as furnaces allow for premium reproduction. Furnace atmosphere can be precisely controlled, and this automation is significant. The consistency and quality that brazing in a belt furnace produces cannot be paralleled.

The brazing process is straightforward. Typically, filler metals have a melting temperature above 450 °C, but their melting point is always less than that of the actual metals being joined. This protects the integrity of the original metals and prevents warping. Once the filler metal reaches its melting point, it flows into the space between the original materials to form an alloy at the brazed joint, as Fig. 1 depicts.

J. Xu et al., *Conveyor Belt Furnace Thermal Processing*,
https://doi.org/10.1007/978-3-319-69730-7_12

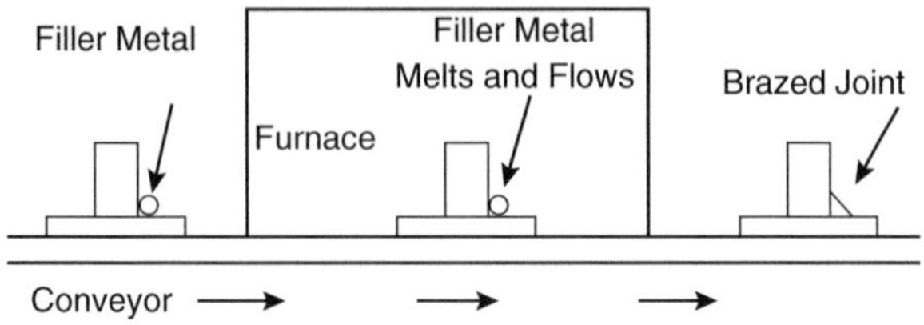

Fig. 1 Schematic for brazing with a belt furnace, as depicted on http://en.wikipedia.org/wiki/Brazing

Parameters of Brazing Process

Because brazing has certain requirements, brazing furnaces must be able to support these parameters. Components get heated, joined, and cooled inside of a furnace, and this entire process is heavily dependent on temperature. The determining factor for the brazing temperature is based on the melting point of the filler metal. Furthermore, temperature control has a strong influence on success. Usually, there are five steps to a typical brazing process cycle: (1) an optional preheat, including holding, (2) ramp to temperature, (3) brazing, (4) cool down, and (5) exit, as illustrated in Fig. 2. During the thermal brazing cycle, the temperature needs to be precisely controlled by the furnace.

Timing is also very important to the brazing process. Brazing requires a minimum duration to ensure that the filler metal can flow through the whole joint to make a strong bond. Generally, 1 or 2 min is enough for this. Too much processing time will yield undesired effects. Typically, the duration of the contact between liquid filler metal and the base metals must be minimized to prevent excess flow of the filler metal. However, in some cases, greater processing time can produce a modified braze alloy that has a higher melting temperature, which could be advantageous in situations where a second brazing may take place.

The atmosphere must be neutral for both the base metal and the braze alloy to avoid oxidation carburizing, decarburization, and nitriding. This ensures that no reactions can occur between the metal, alloy, and the atmospheric elements. In addition to oxygen and carbon, it is possible for nitrogen and hydrogen to also trigger reactions in some circumstances. Likewise, many metals can oxidize rapidly when heated. To avoid this, the brazing atmosphere inside of a furnace must be controlled by utilizing a reducing gas such as hydrogen, forming gas, or cracked ammonia. Controlling the furnace atmosphere is the best way to offset the risk of oxidation and other unwanted issues.

Atmosphere Control in a Belt Furnace

The amount of oxygen and moisture levels in a furnace have a direct influence on the behavior of the filler metal as it melts. For example, when there are low levels of oxygen and moisture in a furnace, the filler metal has a greater ability to melt and

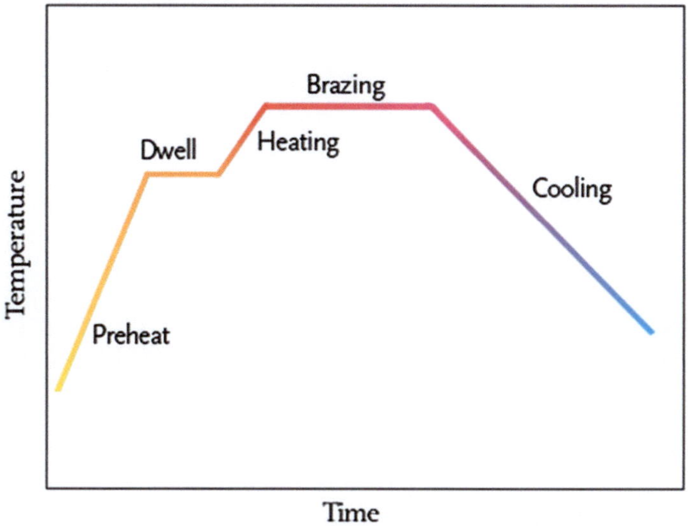

Fig. 2 Schematic of a typical brazing process cycle, as seen in "Furnace Materials #4: Brazing of Metals" by Andersson, Holm, Wiberg, and Astrom

navigate into tight joints and create a stronger bond. High levels of these elements can produce the opposite effect, which is averse to brazing. Maintaining strict control of the furnace atmosphere will significantly improve the quality of the brazing process and prevent physical defects.

Furnace atmospheres contain both neutral and active gases, and these gases can encounter a multitude of reactions. If parts have not been cleaned before entering the furnace, surface contaminants can interact with the gases and cause unexpected results. Additionally, gases can also react with each other, as well as with vapors from the filler metal. Brazing furnaces allow for precise adjustments within the atmosphere composition in order to produce a balance between the two types of gases. This equilibrium helps to control oxidizing and reducing, as well as carburizing and decarburizing. The atmosphere also has an inevitable effect on the rate of heat transfer within a furnace's heating and cooling phases.

It is important for atmosphere control systems in brazing furnaces to include both a system to analyze gas and also a flow and gas mixture control system in order to control the balance of gases and maintain atmospheric precision. Continuous brazing furnaces are specifically advantageous for accurate control, in that different furnace zones can accommodate diverse atmosphere composition set points, as shown in Fig. 3.

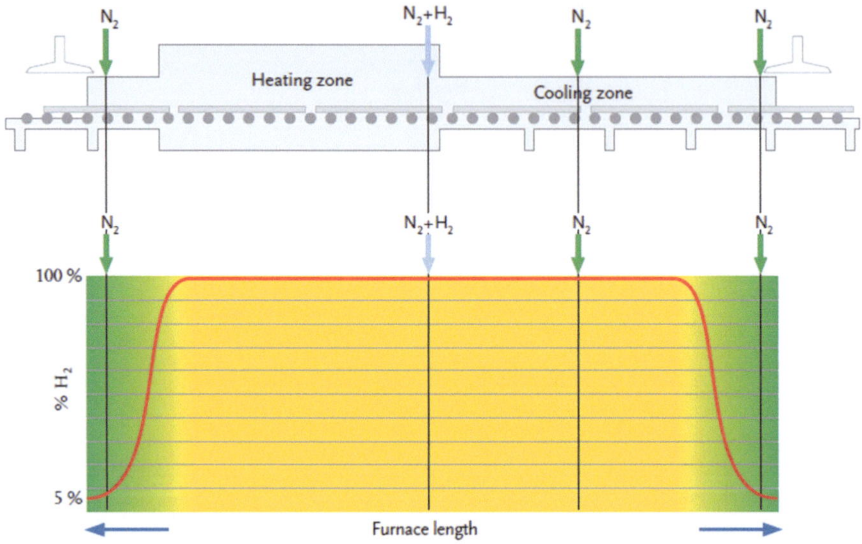

Fig. 3 Atmosphere control in belt furnace, from "Furnace Materials #4: Brazing of Metals" by Andersson, Holm, Wiberg, and Astrom

Selecting a Furnace for Brazing

A high quality furnace is essential for in order to produce top-quality results from brazing. The type of products being brazed, characteristics of the materials, production volumes, and schedule can all influence the furnace that is best suited for the task. A continuous brazing furnace is best for production if a perpetual flow of parts is desired, regardless of whether the parts need to be placed in a tray or basket. A brazing furnace has a metallic muffle, which ensures precise control of the atmosphere while minimizing contamination. This guarantees reliable quality.

The HSA series belt furnace seen below in Fig. 4 is designed to excel at the brazing process. This furnace uses ceramic heater boards to achieve elevated temperatures. The HSA series furnace comes with a refractory heating chamber that is equipped with ceramic fiber FEC (fully enclosed coil) heating board. The heating works to give fast thermal response. The furnace is equipped with a temperature profiling system and a computer monitoring system. Moreover, it can achieve precise atmosphere control for hydrogen, nitrogen, and oxygen by using a dew point and oxygen monitoring system.

Forced air or water cooling is used in the cooling section of the furnace. The muffle is located within the furnace and helps to control atmospheric conditions as

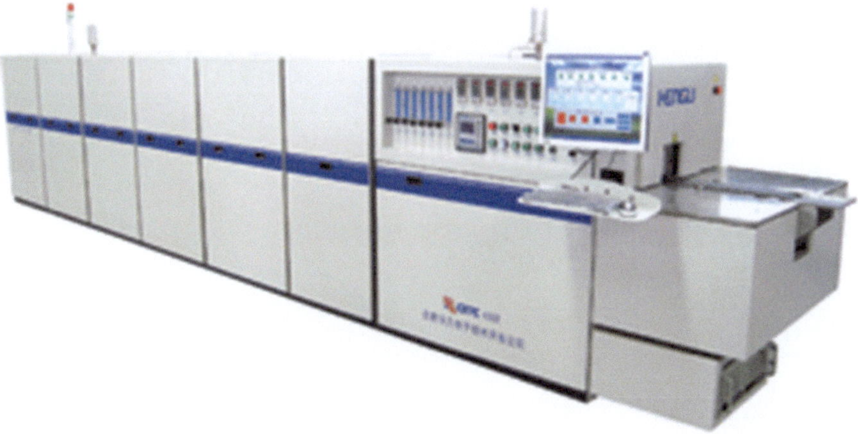

Fig. 4 Hengli HSA series belt furnace

well as to maintain a clean environment inside the furnace. As a standard feature, this furnace is equipped with a steel brush for cleaning the conveyor belt, and ultrasonic belt cleaning is available as an extra option.

The HSA series furnace has a microprocessor-based PID controller to manage the furnace. Type K thermocouples are used in determining the zone temperatures, and controls are located on the right side of the furnace, which can be viewed from the entrance. The central processing unit (CPU) is located at the exit table and is paired with a Windows operating system for ease of use. The computer system is preinstalled with a program for controlling the furnace parameters, including the belt speed, zone temperatures, and atmospheric conditions. Temperature profiles can be stored and retrieved as well for future purposes. The furnace has parameters in the software for capturing, storing, displaying, and printing the furnace profile. Thermocouple ports are located at the entrance table for connecting the profiling thermocouple directly to the microprocessor. This feature allows for monitoring and recording of actual temperatures that are experienced by parts. Additionally, this furnace is equipped with a redundant overheat safety protection system, which incorporates an additional type "K" thermocouple in the center of each controlled zone and the multi-loop alarm. The specifications for a HSA7503-082N belt furnace are shown in Table 1.

Table 1 Specifications of HSA7503-082 N belt furnace

Specification	HSA7503-08ZN
Rate temperature	900 °C max, normal operating: 850 °C
Belt width	30″ (750 mm)
Effective above belt clearance	1″(2Smm)
Control zones	8
Conveyor speed	2–8 IPM (40–200 mm/min)
Loading table	59″ (1500 mm)
Unloading table	59″ (1500 mm)
Belt	Balanced V weave, SUS314
Heating elements	FEC heater board
Insulation	High quality ceramic fiber
Temperature controller	Intelligent PID Shimaden controller
Alarm	Thermocouple, over temp, belt stop. Audio and visual alarm
Atmosphere	6 pipes of dry clean air or N2. 2–6 m^3/h, 1.1–3.3 CFM
Cooling	Forced air cooling
Across belt temperature uniformity	+/–4 °C
Overall system width	60.5″ (1540 mm)
Overall system length	343″ (8705 mm)
Overall system height	54″ (1.350 mm)
Net weight	2500 kg
Power	Three phase, 480 VAC, 60 Hz, 38 KVA MaxNormal operating power draw is about 15 KVA

Conclusion

Brazing in a belt furnace offers an economic opportunity to join complex parts together while creating a high quality bond. Continuous furnaces are extremely versatile, in that they allow for both small- and large-scale operations, as well as reliable results. Maintaining control of the furnace temperature, atmosphere, and processing time is imperative if an efficient brazing process is to be achieved. Precise time control helps to ensure product quality, so that defects are not discovered after brazing. Furthermore, brazing furnaces should offer rapid heating, cooling, and temperature stability in order to produce desired results. Once these requirements are fulfilled, a consistent brazing process can be achieved.

The Influence of Belt Furnace on Post Mold Cure Process

Introduction

Post mold cure (PMC) is one of the most significant processes in electrical industry. This process exposes part of a mold to elevated temperatures in order to speed up the curing process and to maximize some of the material's physical properties.

The PMC process will expedite the cross-linking process and properly align the polymer's molecules to make a stronger part with better high-temperature characteristics. Much like tempering steel, post curing thermosetting can improve physical properties above what the material would normally achieve at room temperature, such as tensile strength and flexural strength, and can modify the temperature of heat distortion. Moreover, post mold cure process is the most common strategy used today for warpage problem solution. Finally, it can also deal with the outgassing phenomenon during IC package.

Given the benefits of PMC processing, it is widely used in the electrical device industry. It is easy to find the application of PMC in many facilities, and there are a large number of companies using PMC on their products. Figure 1 shows various chips that have been packaged using PMC processes.

PMC Process in IC Encapsulation

There are two major manufacturing steps in IC encapsulation industry. For the first step, the IC chip would be encapsulated into a thermosetting epoxy molding compound (EMC), which is the most common plastic material for IC package manufacture today. This pelletized compound is injected into a hot mold die to form the body around the IC die. And after the plastic injection, the mold is cured. The object of this step is to achieve good fill and partial cure of the mold.

J. Xu et al., *Conveyor Belt Furnace Thermal Processing*,
https://doi.org/10.1007/978-3-319-69730-7_13

Fig. 1 Products of PMC

The second step of the process is the post mold cure (PMC) process. The goal of this step is to increase degree of cure and reduce warpage in an IC chip. In order to optimize properties, PMC process provides a critical completion of the cure process to complete the chemical cross-linking of the material. During PMC, the material experiences additional molecular rearrangement and greater efficiency of molecular collisions, thereby resulting in a greater degree of cross-linking. The heating can also cause any residual peroxide to break apart and initiate some additional chains.

Generally, PMC heating process can be divided into three heating stages, assuming that heating process in the furnace is uniform. In the first-stage process, products are heated from room temperature 25 °C to post mold cure temperature 175 °C in a short time. In the second stage, the temperature is held constant for several hours. In the final stage, the product is cooled from 175 °C to room temperature in a short interval.

Applications of PMC

Post mold cure has been applied by many electronics companies, especially semiconductor designers and manufacturers. Table 1 shows various products that have benefited from PMC to achieve better performance as well as some companies associated with them. From the table, it is easy to see that PMC technique is widely used in electronic market by a wide range of corporations. If your company requires better device performance, PMC processing is an easy choice.

Table 1 Some products and companies using PMC process

Products	Companies
IC	Intel; Samsung; Toshiba; SK Hynix; IBM; Sony; AMD; Freescale; Marvell; Nvidia; Qualcomm; ANADIGICS; Cree; Infineon; ST's; Microsemi; Silicon Labs; TI; Vishay; IR; NXP; Intersil; Amkor; Spansion; Renesas
Packaging	ASE; Intel; TSMC; Microsemi; SPIL; QuickLogic; SMIC; UMC; GlobalFoundries; Amkor; UTAC
RF power amplifier	Skyworks; TI; TriQuint; RFMD; Cree; ANADIGICS; Maxim Integrated; Infineon; NXP; ST's; Avago; Semtech; ADI; Linear; MACOM; Freescale; Microsemi; ONsemi; Lattice
Memory	Samsung; Elpida; ISSI; Micron; Maxim; Integrated; Microsemi; SK Hynix; NEC; Panasonic; Toshiba; SPIL; OSE; Spansion; Winbond
FPGA	Xilinx; Altera; Atmel; Microsemi; Lattice; QuickLogic; Actel; Vantls; Cypress
ASIC	LSI Logic; Toshiba
Attenuators	Skyworks; TriQuint; RFMD; Avago; Peregrine; Analog
RF passive	Skyworks; TriQuint; RFMD; PPI; MACOM; Microsemi; ONsemi; RCD
Transistor	TriQuint; NXP; ST's; Avago; ONsemi
Diodes	Skyworks; NXP; ST's; Avago; MACOM; Microsemi; Vishay; ONsemi
Mixers/multipliers	Skyworks; TriQuint; RFMD;NXP; Avago; ADI; Linear; Peregrine; TI; Vishay; Analog
Filters	Skyworks; TriQuint; Avago; MACOM; TI; ONsemi
Modulator/demodulator	Skyworks; TriQuint; RFMD; NXP; ST's; Micron; Avago; Linear; Microsemi; Analog
Switches	Skyworks; RFMD; Infineon; NXP; Maxim Integrated; ST's; Avago; ADI; MACOM; Peregrine; Microsemi; TI; Vishay; ONsemi; Analog
Die/wafer	TSMC; TI; Fairchild; Vishay; IR; Cypress; SPIL; SMIC; Amkor

The Influence of PMC on Material Properties

It is the epoxy molding compound (EMC) that determines the properties resulting from PMC process. Epoxy is the vital part of EMC; it will directly affect the flow characteristics of EMC as well as directly impacting the EMC thermal performance and electrical characteristics. At present there are several commonly used epoxy resin structures and different types of epoxy resin with different characteristics, such as o-cresol-type epoxy resin with high thermal stability and chemical stability, bisphenol A-type epoxy resin with low shrinkage and low-volatile component, and multifunctional-type epoxy resin which has excellent thermal stability, fast curing, and high Tg, and other characteristics such as biphenyl-type epoxy resin which has a low viscosity and high filling characteristics, tea-type epoxy resin with high Tg

Table 2 Some epoxy resins and their characteristics

Epoxy resin	Characteristics
O-cresol-type epoxy resin	High thermal stability and chemical stability
Bisphenol A-type epoxy resin	Low shrinkage and low-volatile component
Multifunctional-type epoxy resin	Excellent thermal stability, fast curing, and high Tg
Biphenyl-type epoxy resin	Low viscosity, high filling
Tea-type epoxy resin	High Tg, high-heat-resistant
Modified epoxy resin	Good flexibility

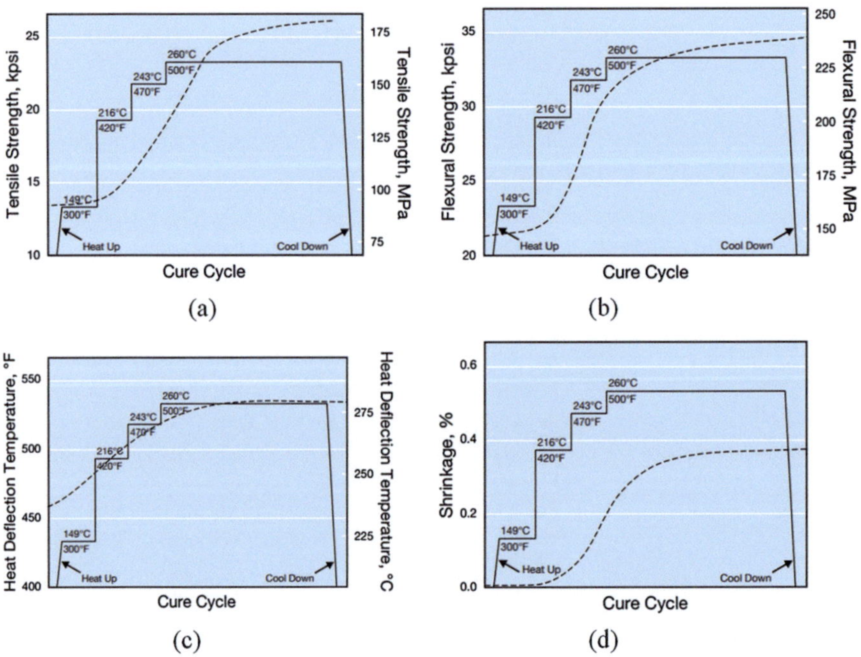

Fig. 2 Physical property changes of cure circle: (**a**) Tensile strength; (**b**) flexural strength; (**c**) heat deflection temperature; (**d**) shrinkage; the specimens are Torlon 4203 L, 3 mm (1/8 in.) thick

and high-heat-resistant properties, and modified epoxy resin with good flexibility and so on (Table 2).

The PMC process involves placing the molded articles in a forced air furnace and thermally treating them to a series of increasing temperatures for various times. The program of times and temperatures is referred to as the cure schedule or cure cycle. During the PMC process, the molecular weight of the polymer increases by chain extension. And as the molecular weight increases virtually, all mechanical, chemical, and thermal properties are affected. Figure 2 illustrates the physical property changes of PMC process. After PMC, the physical properties of objects are substantially increased.

PMC can prevent problems such as warpage during encapsulation in chip packages. In IC encapsulation, one of the prevalent and troublesome EMC defects is warpage. Fortunately, PMC is an efficient method to alleviate the warpage problem during encapsulation. PMC is also one of the principal tools to mitigate outgassing. PMC can remove the volatiles from the cross-linked plastic material. If the volatiles are not removed and the EMC is exposed to elevated temperatures with poor ventilation, one will observe poor in strength, elongation, and compression set properties accompanied by chemical decomposition. Insufficient or poor PMC can result in "smoke," bubbling, delamination, and unsightly sticky surface deposits.

To achieve a satisfactory PMC process, the furnace must be tuned to optimize the cure process, which can only be achieved through high-quality temperature control.

Furnace Selection of PMC

Selecting a suitable furnace requires knowledge of the temperature, time, and atmospheric conditions of the process. Basically, PMC furnace can be divided into batch furnaces and continuous furnaces. Batch furnaces are suitable for any part size but limited with respect to production volume. As batch furnaces use the same door to load and unload the part, these furnaces can only produce one batch at a time. Continuous furnaces are suitable for high-volume production. A continuous furnace uses a conveyor belt to continuously move parts through the furnace. This type of furnace is for high-volume production. Figure 3 shows the Hengli continuous belt furnace.

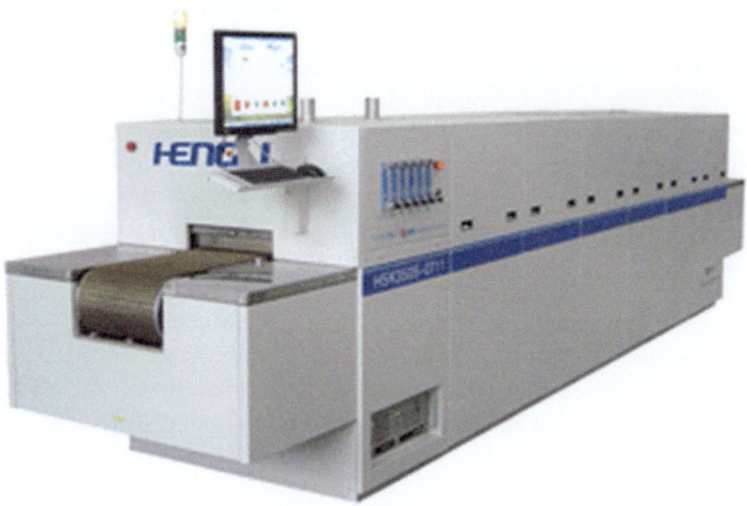

Fig. 3 A Hengli continuous belt furnace

Furnace technology, economics, and part quality influence the decision on whether to use a continuous or a batch operation. The economics questions center around cost of ownership, which can include initial cost, operating costs, repair costs, product yields, and return on investment. Quality issues often are associated with process stability, quality, and consistency, while technology focuses on ease of operation, process definition, thermal cycles, temperature requirements, atmosphere conditions, weight of product, and desired throughput. The questions and their relative importance are always depending on the specific situation.

Most producers are still using batch furnaces for PMC process today, but as many studies and discussions pointed out, it is better to use continuous furnaces for PMC process if you want to get a steady flow of incoming parts. This is because continuous furnaces are extremely versatile and can be employed to perform a multitude of processes. They are an excellent choice for manufacturing medium- and high-volume products. There are a great number of advantages of converting the batch process to continuous, such as:

1. Superior art to part temperature uniformity
2. Increased throughput
3. Process combination
4. Lower up-front investment
5. Reduced changeover times
6. Part loading flexibility

A continuous furnace is ideal for processes requiring high production volumes, process consistency, and precision control. All components can go through the furnace smoothly. And during the furnace process, as the continuous furnace can serve consistent heating process, the consistency of products could be ensured in a high level. The defects will also be effectively prevented and eliminated. Furthermore, a continuous furnace can greatly improve the production efficiency by being continuously available, rather than intermittently (as is the case for the batch furnace, which must heat up and cool down). In addition, a good continuous furnace is often much more compact than a batch furnace, which is beneficial for floor space considerations and facility costs. Moreover, a continuous furnace is easily used for automation offers.

When choosing continuous furnace, the air convection heating function should be considered, too. Unlike the traditional furnace using radiative heating, a hot air convection furnace can elevate the temperature through convective heating which offers an extreme uniformity to PMC process. Figure 4 illustrated the difference between traditional radiative heating and convective heating. More important, the hot air convection furnace can get high energy efficiency up to 70% saving over conventional ovens. Additionally, convection furnace is always environmentally friendly with no noise and pollution operation.

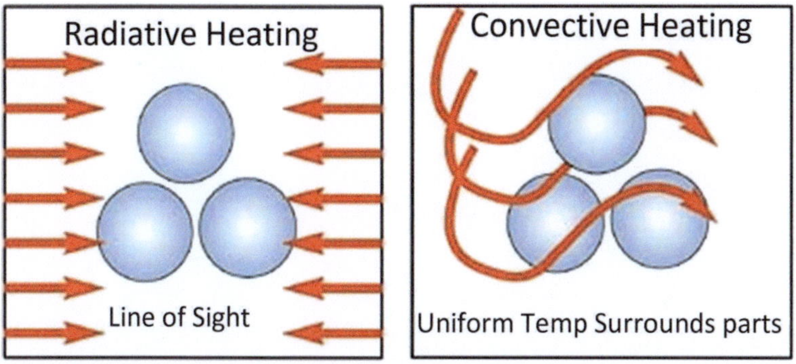

Fig. 4 Radiative heating process and convective heating process

Furnace Control of PMC

For most PMC process, the longer cure profiles require longer furnaces. These furnaces are more suitable for the inline, integrated manufacturing line used by many printed circuit board assemblers. The PMC process needs specific control of temperature and time, which is critical for getting excellent performance after PMC process. If the temperature of furnace during PMC is higher than the setting point, it is easy to damage the components which cause the failure of production. And if the furnace temperature cannot reach the required point, the post curing would be insufficient, which will lead to the great reduction of PMC quality. Consequently, it is vital for furnace temperature control to achieve high-quality PMC performance.

As the PMC process is sensitive for the temperature, uniform furnace temperatures are essential for the PMC process. For most cases, the greatest difference in temperature between the hottest and coldest point of the oven can be tolerated. Generally, a hot point occurs near the air intake while a cold point near the exhaust vent.

Controllers programmed to raise the temperature by 0.3 °C (0.5 °F) per minute are recommended. Automatic shutoff and manual reset features are also desirable. A good oven is supposed to cut off automatically when the temperature reaches 2.8 °C (5 °F) above the set point. This is required to avoid distortion of the parts which can occur if the temperature exceeds the deflection temperature of the part.

Belt Furnace for PMC

The HSF series hot air convection furnace is an efficient belt furnace designed and used for post mold cure process. This furnace can make temperature to 400 °C. It can be heated by infrared and/or hot air circulation heating depending on process parameters and requirements, and the temperature profiles are able to run at the

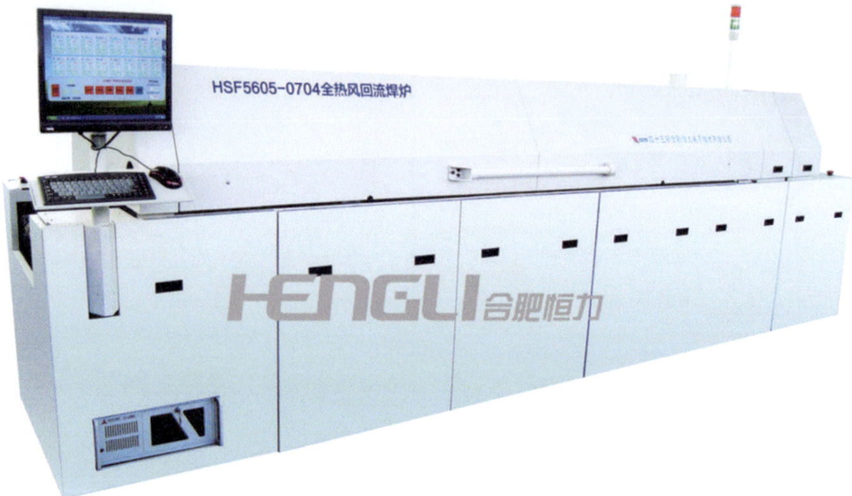

Fig. 5 A Hengli HSF series hot air convection belt furnace

desired heating rate to meet the required PMC temperatures under controlled atmosphere. Its air or nitrogen atmosphere can serve a completed curing process. Temperature control zones offer precise control which allows the furnace to run at the proper heating rate to meet the needs for curing. The HSF series belt furnace can offer uniform temperature distribution to meet the qualification of PMC process. Its conveyor system allows proper heating across the belt with little temperature variations. And the HSF series furnace comes with an ultraclean heating chamber, which can give rapid thermal response. Figure 5 shows the HSF series furnace.

The furnace is long enough to handle PMC process. To ensure proper practice for continued use, technical information and training will be given upon installation of the furnace. A microprocessor-based PID controller provides appropriate system control. Type K thermocouples are used in determining the zone temperatures. The central processing unit (CPU) is located at the entrance table and is available with a windows operating system for ease of use, and the program is installed ready to control furnace parameters such as belt speed, zone temperatures, and atmospheric conditions. Temperature profiles can be stored and retrieved as well for future purposes. There are programs for capturing/storing, displaying, and printing out the furnace profile which is already included in the software. Additionally, the furnace is equipped with a redundant overheat safety protection system which incorporates an additional type "K" thermocouple in the center of each controlled zone and the multi-loop alarm.

Conclusion

PMC is an important technology for electrical industry. It can highly improve the properties of chips and is widely used by a great number of companies. A large number of electronic products and companies have used PMC for better performance. The quality of PMC process is deeply influenced by the temperature and time, which are strongly influenced by the furnace. A good continuous belt furnace with precise control and convective heating will offer great conditions for PMC process.

An Introduction to Glass-to-Metal Seals

Introduction

A true glass-to-metal seal is defined as one in which inorganic glass is heated to the point when intimate contact ("wetting") is attained upon a hot metal surface and is retained when the glass and metal cool to room temperature. Typically, the glass and metal components share similar rates and coefficients of expansion.

The glass layer is electrically insulated and its resistance is dependent upon the glass type, temperature, and surface condition. However, the glass can be given a metallic oxide coating to conduct electricity. Glass can perform under high pressure, vacuum, high temperature, thermal cycling, radiation, shock, and vibration and will not degrade with time.

Types of Glass-to-Metal Seals

There are diverse forms of glass-to-metal seals, and they can be categorized by their definitive structures:

Internal seals are those in which the metal, in the form of wire, rod, ribbon, or tube, is surrounded by glass.

External seals are those in which a band, tube, or cup of metal adheres to the edge of a glass insert, which may be a round button, disc, square, or any other shape.

Because the glass is much more vigorous in compression than in tension, the difference in the coefficients of expansion for glass and metal components may be somewhat greater in external than in internal seals. Where expansion mismatch occurs, the ratio between the outside diameter of the tube in an external seal and its wall thickness should not be less than 15:1, with a maximum wall thickness of 0.01 in. Where expansion coefficient differentials are minuscule, as in Kovar-to-metal seals, these restrictions do not apply.

J. Xu et al., *Conveyor Belt Furnace Thermal Processing*,
https://doi.org/10.1007/978-3-319-69730-7_14

Tubular seals are "inside tubular" when the metals are surrounded by the glass and "outside tubular" when the glass is surrounded by metal.

Tubular seals are often called "housekeeper" seals. Having a considerably contrasting expansion coefficient, the metal, such as copper or stainless steel, is formed to a thin edge. This enables it to yield as it expands with heat relative to the glass so that the stress of the glass does not exceed its tensile strength.

Edge seals resemble tubular seals in that they may be either normal edge or a "housekeeper" type (geometry and applications differ). Edge seals are those in which the edge of the metal is embedded in the glass.

Butt seals are produced by sealing a metal disc, or ring, to the end of a piece of glass tubing.

Window seals may take on any shape; however, round windows are subject to less severe and complex stresses. Combinations of different types of seals are also possible.

A more common way to categorize seals is by their thermal expansion coefficient (CTE). Using the classification system based on relative thermal expansions, systems can be described as either *matched* or *unmatched*. The classification can further be divided into subcategories based on the type of glass, type of metal, and how the two are compounded together (Fig. 1).

Matched seals are subdivided according to the CTE of the glass. Glass with CTE lower than 6×10^{-6}/°C is classified as hard glass (e.g., borosilicates and vitreous silica). Glass with CTE higher than 6×10^{-6}/°C (e.g., soda glass) is classified as soft glass.

Unmatched seals are divided into two primary groups: ductile and compression. Ductile seals (the housekeeper is the best-known example) are most commonly used for combining glass and copper. Compression seals form a mechanical bond by establishing a compressive hoop force from the metal onto the glass (shown schematically in Fig. 2).

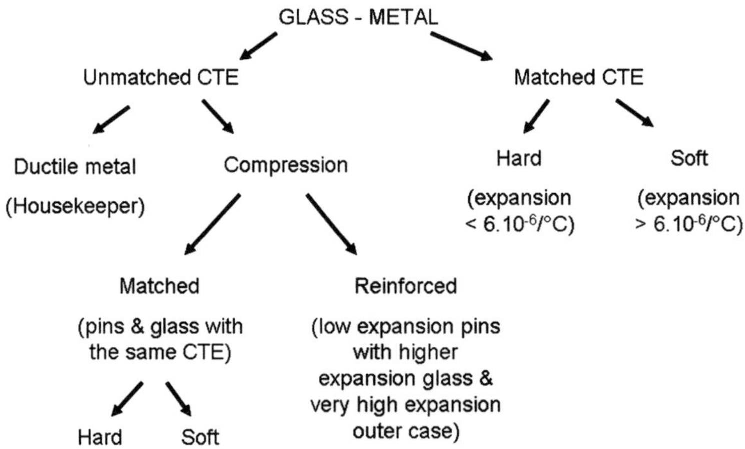

Fig. 1 Categorization of glass-to-metal seals (From TWI)

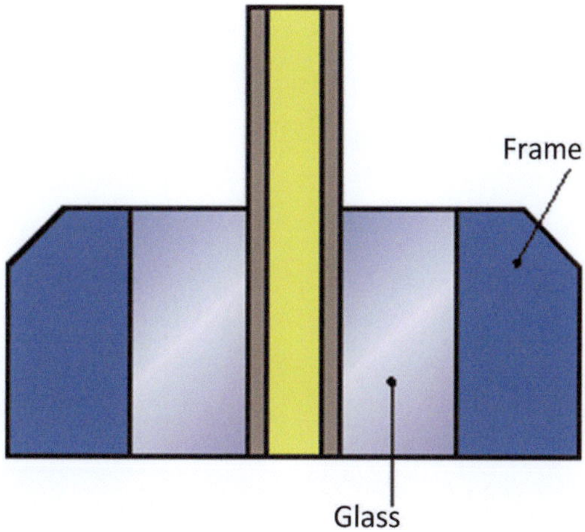

Fig. 2 Schematic of compression seals (From TWI)

Furthermore, there are two main subdivisions in compression seals: *matched* and *reinforced*. Matched compression seals have pins and glass with the same CTE value. The compressive force is generated by the outer case (frame) with a considerably higher expansion. Matched compression seals can be even further divided by specifying the glass type (hard or soft). Reinforced compression seals use low expansion pins with high expansion glass and a profoundly high expansion outer case.

Windows and combination seals are excluded; soft glass has been used for all types of internal, external, tubular, and edge seals. Hard glass discovered that its greatest application is in internal and tubular seals; however, other classifications are also possible.

A perfect CTE match is not usually attainable. This can be due to the fact that generally the elongation of a piece of metal is in a linear relation with respect to temperature, while the glass is not (shown in Fig. 3). Therefore, the goal is to achieve a relatively contiguous value between materials.

Thermal hysteresis is conveyed as existing in glass, so it is essential that all glass-to-metal seals are heated and cooled slowly. The magnitude of the strain is determined by the difference in CTE between the glass and the metal.

Fabrication of Glass-to-Metal Seals

In most glass-to-metal seals, the procedure includes:

1. Oxide formations on metal surfaces by heating to the proper temperature
2. "Wet" the oxides by fused glass to form a glass-oxide-metal transition
3. Seal the metal-glass bond

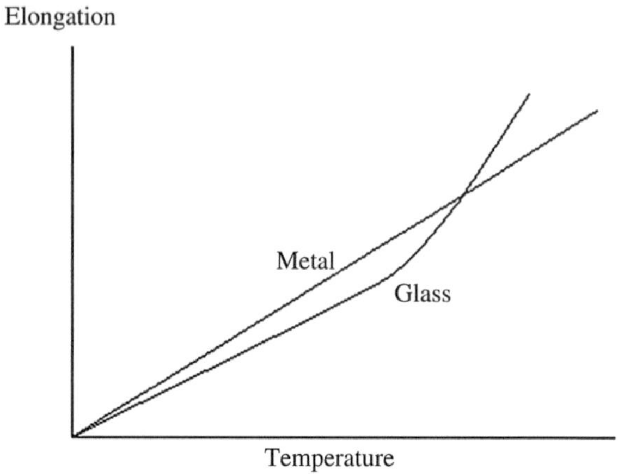

Fig. 3 Typical forms of glass and metal thermal expansion curves (From *Handbook of Electron Tube and Vacuum Techniques*)

It is worth noting that a reliable hermetic seal can be produced without the presence of oxides (e.g., platinum).

Housekeeper seals can be formed with any kind of glass (usually soft glass) if the materials applied fulfill the following requirements:

1. The expansion of the metal portion must be larger than that of the glass in order to assure that the stresses normal to the glass-metal interface will always be compressive.
2. The metal being used must be soft and thin enough to be able to deform plastically in order to allow dimensional changes that take place in the glass as it cools.
3. The metal must form a durable bond with the glass.
4. The dimensions and shape of the seal must be designed to provide a large contact area between the metal and the glass while, at the same time, avoiding any tensile stress perpendicular to the surface or from any other direction.

Copper can fulfill all these requirements; however, platinum, iron (steal), and molybdenum can also be used for such seals. In using these metals, the seal can be constructed into a variety of shapes: wire seal, ribbon seal, feather-edge seal, or a disc seal.

Copper-to-Glass Seals (Fig. 4)

Although the CTE difference between glass and copper is quite extensive, a satisfactory vacuum seal can be constructed which utilizes the property of glass to "wet" copper oxide, some of which diffuses into the glass and forms a transition phase. For a satisfactory seal, the oxide layer should be no more than a few tenths of a micron thick and consist of various oxides.

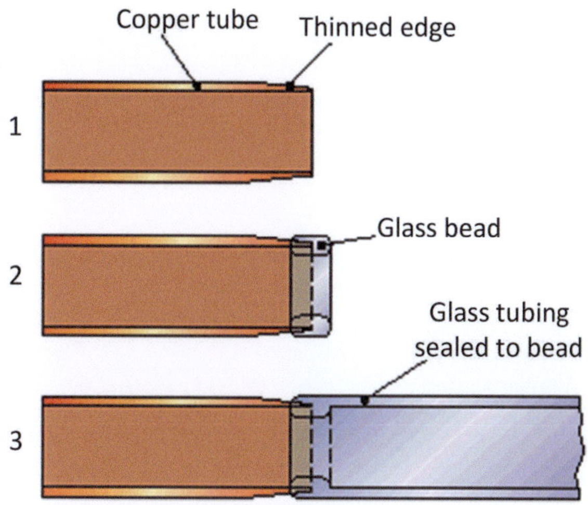

Fig. 4 Housekeeper copper-glass seal (From TWI)

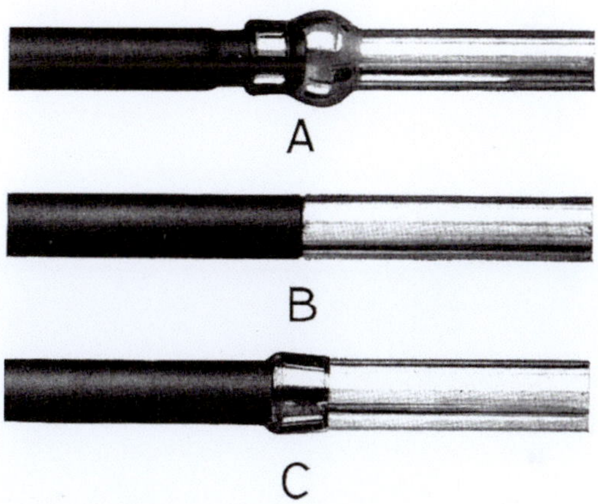

Fig. 5 Various types of housekeeper copper-glass seals (From *Handbook of Electron Tube and Vacuum Techniques*)

1. The copper piece is tapered down to a very thin feather-edge seal in order to follow the expansion and contraction of the glass. The feather-edge seal must be polished and it must be smooth.
2. Degreasing and hydrogen firing at 800 °C.
3. The copper piece is usually borated by reddening it from heat in air and quenching it in a concentrated solution of sodium borate. The color of the copper at the seal site should have a uniform deep red or purple sheen.
4. The copper and glass are heated to 1000 °C in air and then brought together. This process is called *beading*.
5. Copper-glass seals must be carefully annealed in an oven immediately after glassing.
6. The assembly is chemically cleaned to remove the excess copper oxides.

Figure 5 shows the various types of housekeeper copper-glass seals.

Summary

The advantages and disadvantages are summarized in the table below.

Advantages	Disadvantages
Extremely high durability	Unable to cope with large CTE mismatch
Mechanical stress and strain resistance	Precise control of glass composition needed
High shock and vibration resistance	
Thermal shock stability and resistance	
Chemical stability and resistance	
Absolute hermetic	
Inexpensive, easy to manufacture	
Electrical insulation	

Furnaces for Glass-to-Metal Seal Applications

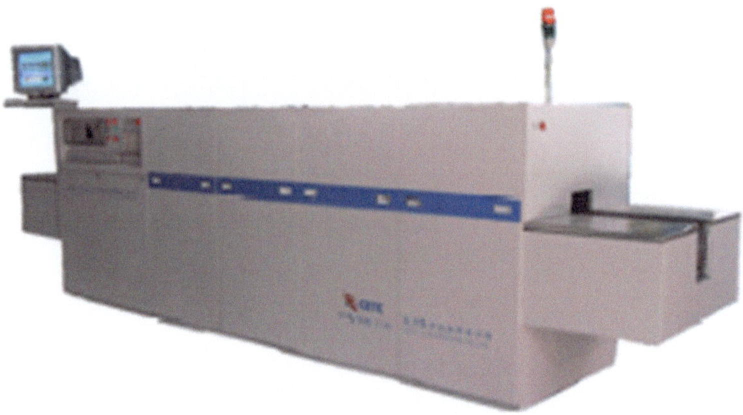

HSA1310-0611NH

A continuous conveyor furnace is used for annealing/brazing applications with an N2 and/or H2 protective environment. The unit features 3150 mm (124″) of heated length comprised of six independently controlled heat zones. Process materials are carried through the furnace on a 125-mm (5 in.)-wide belt with 75 mm (3 in.) of product clearance over the belt.

Controls are located on the right-hand side of the furnace (as viewed from the entrance) and are controlled by a rugged industrial computer.

A rugged industrial computer control system consists of a flat panel VGA color graphic monitor, industrial computer, and Windows based furnace control software for full temperature, conveyor speed, monitoring and setting of profiling, and recipe storage (option if selected). The CPU and monitor are mounted on a swing arm at the exit end of the furnace. Each furnace zone is monitored and controlled using a type "K" thermocouple in the center of each heated zone. These ports connect to the microprocessor, described above so that profiling thermocouples can be connected and used with the software to capture, display, print out, and store profiles. Product cooling is accomplished by utilizing cooling water.

The furnace is equipped with an Inconet/SOS muffle, pilot lights to burn off unburned excess hydrogen, and it is capable of handling hydrogen and nitrogen atmospheres. An oxygen analyzer and a dew point monitor can be built into the system as an option as well.

Furnace Specification

Maximum temperature rating	1100 °C max
Operating temperature range	Ambient 1070 °C
Cross-belt temperature uniformity	±2 °C typical at high-temp zones
Total controlled zones	Six controlled zones
Working dimensions	125 mm (5″) wide × 75 mm (3″) of product clearance over the belt
Belt	Nichrome V mesh
	Balanced weave
Belt speed range	25–150 mm per minute/1–6 IPM
Heated length	3150 mm/124″
Overall length	7000 mm/275″

Furnace Layout

Specification	HSA 1310-0611NH
Loading table	600 mm/24″
Preheating	750 mm/30″
Heating zone	2400 mm/96″

Specification	HSA 1310-0611NH
Water cooling	1300 mm/52″
Transition	450 mm/18″
Unloading table	600 mm/24″
Zones	6
Total length	7600 mm/299″
Power	32 kW max
	12 kW typical
Belt width	125/5″
Belt speed	25–125 mm/min
	1–5″/min
Belt loading	40/8
Opening height	75/3″

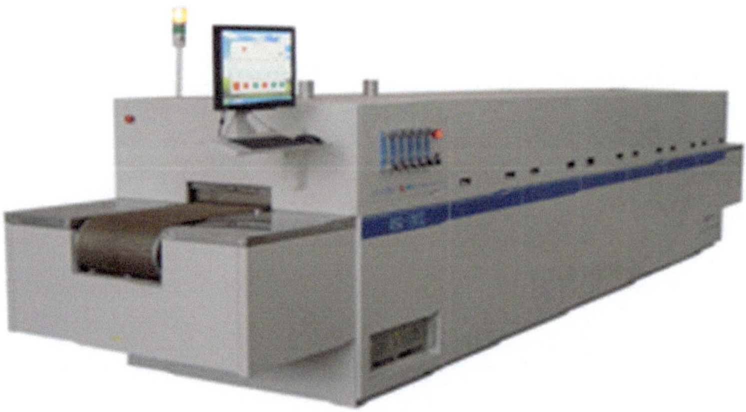

HSA 1503-0409NH

Advantages

Rated to 850°C, the HSA controlled atmosphere series features an ultraclean low-mass belt furnace. The ability to switch atmosphere from dry air to nitrogen or hydrogen and the wide range of belt speeds allow the furnace to be used for both curing and firing operations.

1. Six-channel temperature profiler unit for independent temperature profiling which includes 3 T.C., sampling unit and analyses software, LCD data display and check, and RS232 interface to computer (this is an option; extra cost applies).
2. Atmosphere distribution and management system eliminate thermal shock and process contamination.
3. Ultrasonic belt cleaning system including drying system (this is an option; extra cost applies).
4. Windows DSC-based profiling and monitoring system for monitoring and recording the firing process, including temperature and speed.
5. UPS for the computer system and conveyor drive for power failure (this is an option; extra cost applies).
6. Extraction of burn-off effluents across entire chamber width improves yields.

The furnace is equipped with an Inconet/SOS muffle, pilot lights to burn off unburned excess hydrogen, and it is capable of handling hydrogen and nitrogen atmospheres. An oxygen analyzer and a dew point monitor can be built into the system as an option as well.

Main Characteristics

Specification	HAS 1503-0409NH
Rate temperature	850 °C/normal operating temp: 200–1000 °C
Belt width	150 mm/6″
Above belt clearance	75 mm/3″
Heating length	1200 mm/47″
Insulation cooling length	100 mm/4″
Water cooling length	485 mm/19″
Air cooling length	100 mm/4″
Control zones	4
Conveyor speed	75–500 mm (3–20″)/min
Heating elements	Ceramic FEC heating board
Insulation material	High-quality ceramic fiber
Overall system width	1050 mm (42″)
Overall system length	4260 mm (14 ft)
Overall system height	1350 mm (53″)
Typical temperature uniformity	±2 °C
Controlled atmosphere	Nitrogen is standard, hydrogen as required
Power	480 V, 3 phase, 5 wire, 60 Hz, 16kw/heat insulation power draw: 6 kw

References

1. Fred, R. (1965). *Handbook of electron tube and vacuum techniques*. Boston: Addison-Wesley Publishing Company.
2. Hull, A.W., & Burger, E.E. (1934). Glass-to-metal seals. *Journal of Applied Physics, 5*, 384.
3. Monack, A. J. (1946). Theory and Practice of Glass-Metal Seals, Part III. *Glass Industry, 27*(10), 502–504.

Furnace Temperature and Atmosphere Influences on Producing Lithium Iron Phosphate ($LiFePO_4$) Powders for Lithium-Ion Batteries

Introduction

Discovered by John Goodenough at the University of Texas in 1996, lithium iron phosphate is still advancing as a technology that is utilized in computers, cell phones, and hybrid vehicles. A theoretical discharge process with cathode and anode material inside a battery is shown in Fig. 1.

The cathode material of lithium iron phosphate is valued for its high theoretical capacity, cycle/thermal stability, and environmental benefits over other Li-ion-type batteries. Other materials used for cathode powder material are lithium cobalt ($LiCoO_2$), lithium nickel ($LiNiO_2$), and lithium manganese ($LiMn_2O$). The structure of $LiFePO_4$ is also shown to have an olivine structure providing a safer charging/discharging than other cathode materials [1]. This material shows good theoretical capacity at $170\frac{\text{mAh}}{g}$ and a flat voltage of 3.4 V, but low conductivity is the major disadvantage of these types of powders. Using additives in the form of a carbon source such as acetylene black or sucrose increases electronic conductivity of the powders almost to the theoretical capacity [1]. Carbon coating has also been shown to be a way to decrease particle size and in turn improve electronic conductivity [2]. For example, Wang et al. obtained carbon-coated nanoparticles of LFP/C that delivered a capacity of $134.2\frac{\text{mAh}}{g}$ at a rating of 1 C. Even though this capacity is lower, the cycling ability and the charge losses were minimal which would be well suited for long-term use. Firing temperature and atmospheric conditions are important because every process requires a heat treatment step. For example, a glass-type LFP production route is shown to have improved performance when firing in 7% H_2/N_2 rather than air. There are two main categories of techniques involved in creating $LiFePO_4$: solid state-based and solution-based methods. These methods differ in the technique, temperature, heating rate, and mixture of components.

J. Xu et al., *Conveyor Belt Furnace Thermal Processing*,
https://doi.org/10.1007/978-3-319-69730-7_15

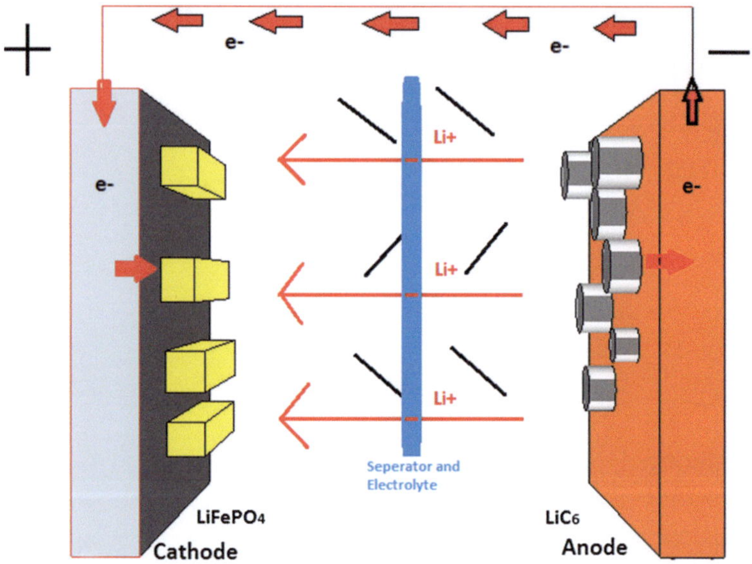

Fig. 1 Battery representation and lithium iron phosphate as cathode

Solid State Processing Methods

Solid state reaction methods include solid state synthesis, carbothermal reduction, and mechanochemical activation among others. The process of solid state synthesis needs much higher temperatures and longer sintering times than other techniques, but it is able to create an ordered crystal structure in a simplistic way compared to solution methods. This technique is well suited for the mass production of $LiFePO_4$ and other unique forms such as ceramics or piezoelectrics [3]. The process of solid state synthesis involves the mixing of precursors followed by heat treatment. Common examples of precursor components used in solid state synthesis for mixing are Li_2CO_3, $FeC_2O_4{\cdot}2H_2O$, and $NH_4H_2PO_4$. Atmospheric conditions used are air, argon, nitrogen, or a combination hydrogen/nitrogen. Adding a conductive carbon coating such as graphite or acetylene black to the precursors will be able to increase the conductivity of $LiFePO_4$ as well. During ball milling and mixing, a stoichiometric amount of precursors is stirred together and milled for 3–8 h. This is followed by heat treatment using equipment such as belt furnaces with temperatures seen up to 700 °C. Heating up to high temperatures can be done in a one- or two-step process. In the two-step process, pre-calcination (pre-firing) is done first involving heat treatment between 250 and 350 °C to expel gases and decomposing of precursors [2]. The final step is to bring the product to the final calcination temperature ranging from 400 to 800 °C [3]. A simple outline of procedures for the popular solid state synthesis is shown in the image below (Fig. 2).

Many different combinations of temperature and heating rates have been studied to find the most optimal outcome. Yamada et al. reports that the highest discharge

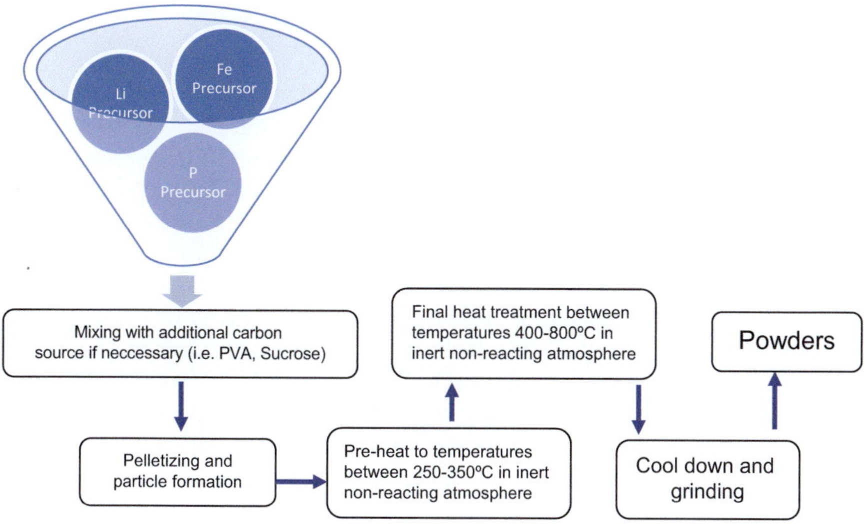

Fig. 2 Solid state synthesis method

capacity found was 162 $\frac{\text{mAh}}{\text{g}}$ at sintering temperatures between 500 and 600 °C with particle sizes below 30 nm. Impurities can appear during the heat treatment phase because of the $FePO_4$ precursor inability to convert fully to $LiFePO_4$ as seen in studies by Yang et al. [4]. The addition of glucose as a carbon source helps inhibit particle growth during sintering therefore increasing conductivity [4]. Solid state methods give a simple way to create LFP powder with the ability to easily control results. It does not require solvent which can be expensive on an industrial scale. Solution-based methods require this solvent but have the advantage of obtaining better results.

A source of carbon is usually added to solve the problem of poor electrical conductivity. For example, Yamada et al. prepared asphalt powder of 85.37% carbon content mixed with LiOH and $FePO_4$ in a solid state reaction. The measured discharge capacity was 138.2 mAh/g at 0.5 °C rate after 50 cycles. $LiFePO_4$/C with carbon derived from glucose shows smaller crystallite sizes of 25 nm as opposed to bare $LiFePO_4$ with crystal size of 32 nm for [6]. The smaller sizes for $LiFePO_4$/C can be attributed to a lower degree of crystallization. Carbon inhibits particle growth which leads to better electrochemical performance [4]. $LiFePO_4$/C has better discharge retention of 100.1% as opposed to bare LiFePO4 that has 52.4% meaning better cycling performance and lifespan for the battery.

Solution-Based Processing Methods

$LiFePO_4$ cathodes by solution-based methods show higher discharge performance, smaller particle sizes, and more homogenous carbon coating. They also require less time and energy input than solid state reactions. Techniques in this group include hydrothermal, sol-gel, and spray pyrolysis. Coprecipitation is one such method that involves mixing of lithium and phosphate compounds with control of pH values followed by precipitation and heat treatment. This process is able to obtain small particle sizes and good uniformity. According to Jugovic et al., crystallite sizes range from 56 to 140 nm and give a maximum discharge capacity of 160 mAh/g. In this method, the process begins with the mixing of lithium and phosphate precursors in solvent. Common elements of chemicals used are Li_2CO_3, $FeSO_4{\cdot}7H_2O$, and H_3PO_4 mixed in deionized water. The mixture is then precipitated and filtered under an inert atmosphere. Afterwards, heat treatment is performed in a furnace with temperatures ranging from 500 to 800 °C under a N_2 atmosphere for times as long as 12 h [3]. A simple outline of the process is shown below (Fig. 3).

Coprecipitation can also be combined with other processes such as solid state synthesis discussed before. This is performed by using a two-step drying process at 80 °C and a high-pressure filtering method. Powders were obtained with 7 wt% carbon and a gravimetric discharge capacity of 167 mAh/g [4]. This value is nearly the theoretical discharge value of $LiFePO_4$ due to the well-homogenized carbon content in the precipitate.

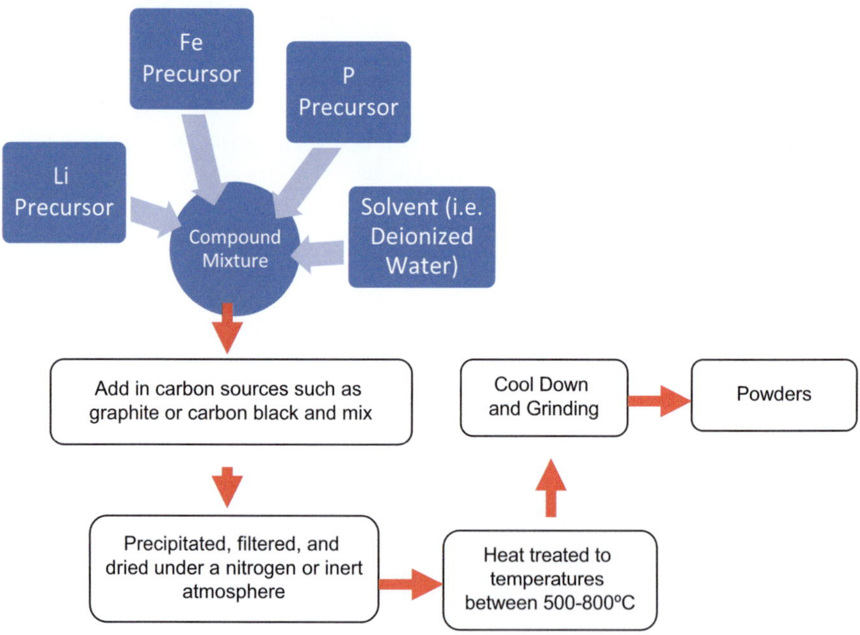

Fig. 3 Solution-based coprecipitation process method

Firing Temperature Influence on Powder Formation

All mechanisms of creating $LiFePO_4$ will require some kind of heat treatment as a final step. Temperature has a significant influence on resulting physical characteristics and electrical properties. High temperatures in solid state reactions are needed in large-scale commercial production of $LiFePO_4$ for repeatability in the manufacturing process. In a spontaneous precipitation method, Wang et al. [2] created LiFeP/C nanoparticles by heating products to 350 °C for 5 h to rid the particles of excess gasses. The products are then heated to 700 °C under a N_2/H_2 atmosphere. Temperature was found to be an important factor in the development of the particle growth. Yamada et al. [6] found that the highest discharge capacities obtained were heat treated between 500 and 600 °C using a solid state synthesis method. These samples gave discharge capacities of 162 mAh/g. Temperatures higher than 600 °C create uncontrollable particle growth, while temperatures lower than 500 °C create unwanted oxidation of unstable Fe^{2+} ions [5].

In coprecipitation, heat treatment up to 800 °C can be used. Firing temperatures of 600, 700, and 800 °C are studied by Jugovic et al. to compare the properties obtained at each firing temperature. Samples obtained at 600 °C were the smallest in size but showed high agglomeration as seen in Fig. 4. At 700 °C the morphology of the compounds does not have any significant change. At 800 °C the morphology of the product changes dramatically creating more crystallization and growth of particles [6].

Seven hundred degree Celsius is the optimum temperature for an aqueous precipitation method because of good crystallinity, small powder particles, and exceptional iron ordering [6]. Depending on the homogeneity of the carbon content and the precursor ingredients, temperature control is of importance.

In another solution-based method called sol-gel, high temperatures play an important role in the curing process. Liu et al. created a mixture containing lithium hydroxide monohydrate, ferric nitrate, and phosphoric acid, with added ascorbic acid as carbon source. It was prepared and mixed in different annealing temperatures of 500, 600, 700, and 800 °C. The heating rate was at 2 °C/min. An increase in crystallinity and grain size has been shown with a rise in temperature. Crystallite sizes obtained from experiments by Liu et al. are shown in Table 1.

The structure can be controlled through the annealing process. The optimum temperature for heat treatment was found to be 600 °C. At this temperature, the initial discharge capacity was 312 mAh/g for the initial cycle and stays at 218 mAh/g after 20 cycles [3]. A higher performance seen here is due to the large surface to volume ratio and poor crystallinity which provides a less packed structure leaving space for more lithium-ion intercalation [3]. The comparison of various annealing temperatures can be seen in Fig. 5.

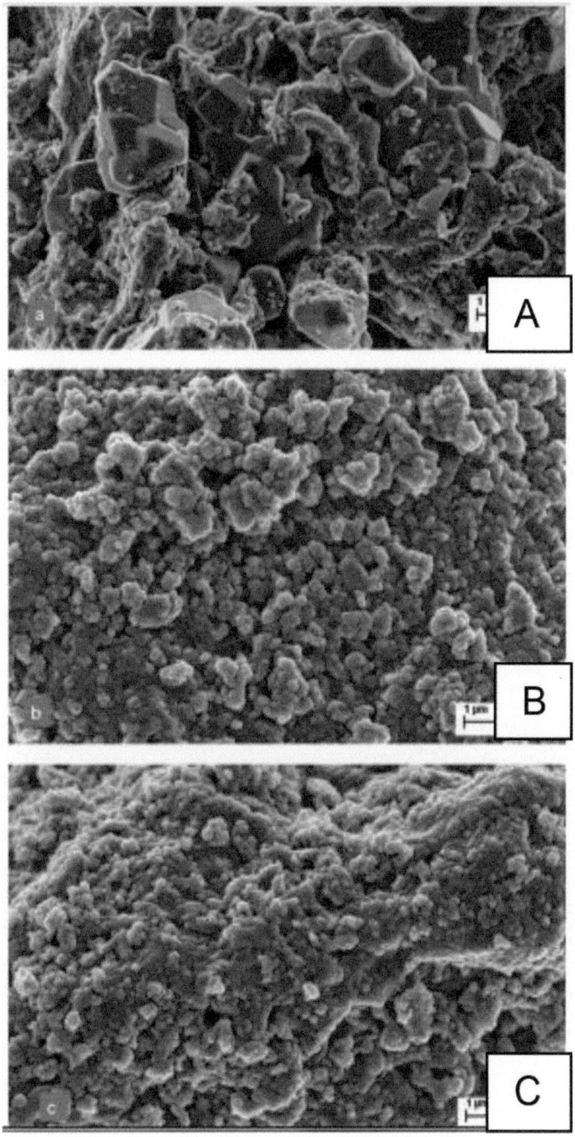

Fig. 4 SEM micrographs of powder obtained from firing at (**a**) 600 °C, (**b**) 700 °C, and (**c**) 800 °C [7]

Table 1 Temperature in relation to crystallite sizes

Temperature (°C)	Crystallite (nm)
500	16.1
600	19.4
700	21.1
800	30.8

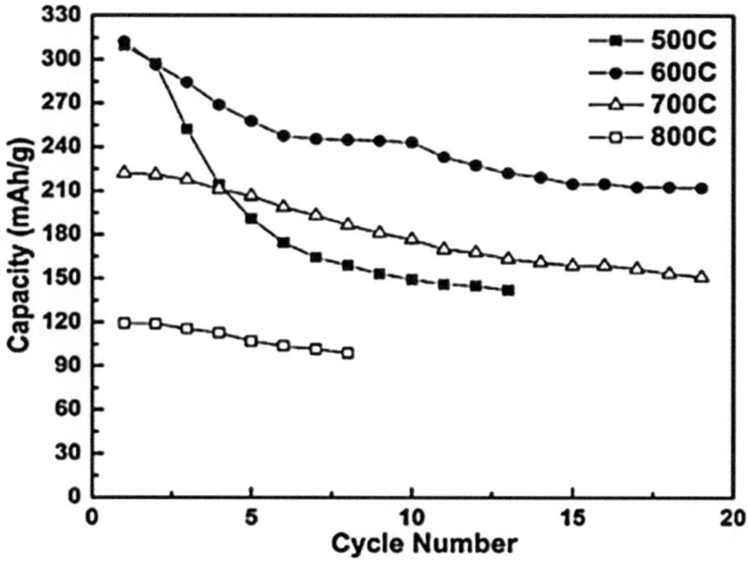

Fig. 5 Number of cycles vs. capacity at different annealing temperatures

Influence of Atmospheric Conditions on Characteristics of $LiFePO_4$

Atmospheric conditions play an essential role during annealing. Controlling heat treatment atmospheres is important in avoiding oxidation of chemical compounds such as Fe_2O_3 or $LiFe_2(PO_4)_3$ [2]. Atmospheres used include air, nitrogen, argon, argon/hydrogen, and argon/nitrogen. Negamine et al. [7] found that hydrogen gas in atmosphere was shown to have influence on iron ions especially in the small particle size range of less than 2 um. Molar fractions of $LiFePO_4$ are important in determining the final makeup of compounds. As seen in Table 2, air results in a molar fraction of 25 mol%, while 7% H_2/Ar shows a much higher product efficiency of 49 mol% [7].

In this experiment, products were heat treated between 400 and 600 °C at its crystallization peak temperature with a heating rate of 10 °C/min. The heating in air shows much less efficiency because of the oxidation of Fe^{2+} ions taking place. Reduction of Fe^{3+} to Fe^{2+} ions takes place in the H_2/Ar atmospheres enhancing the formation of $LiFePO_4$.

Table 2 Molar weight percent in different atmospheres

		$XLifePo_4$ (mol%)		
Particle size (μm)	Quenching type	Air	Ar	7% H_2/Ar
63–40	Rapidly	25	49	49
<2	Rapidly	4	72	81
63–40	Slowly	—	—	58

Furnace for $LiFePO_4$ Manufacturing

The HSA series furnace is an efficient furnace designed and used for heat treatment for processes such as $LiFePO_4$. It features a low-mass refractory heating chamber equipped with ceramic FEC (fully enclosed coil) heating board. The HSA furnace has six independent zones able to reach 1150 °C, temperatures high enough to heat treat lithium iron phosphate. The temperature profiles are able to run at the desired heating rate to meet the required sintering temperatures under controlled atmosphere. Gas is independently controlled for the adjustment of air, nitrogen, hydrogen, and argon inflow. Temperature control zones allow the furnace to run at the proper heating rate to meet the needs for firing and curing. The conveyor system allows proper heating across the belt with little temperature variations. The furnace is monitored by type K thermocouples with each zone monitored by single loop PID temperature controllers. This enables precise and stable temperature control throughout the heating process to the cool down.

The furnace is large enough to handle manufacturing processes for $LiFePO_4$ applications. It has a 14″ (35 cm) belt width and a 130″ (330 cm) heated length. The conveyor system can run at speeds of 1–8 in./min. The HSA furnace is protected from overheating, over loading, and low gas pressures. It also comes equipped with antishock protection on the doors during maintenance. Technical information and training will be given upon installation of the furnace to ensure proper practice for continued use. Emergency buttons and removable collection traps are also located on each end of the furnace. For the cleaning system, there is a rotating motor metal brush that cleans the belt so future products will not be contaminated. Below is a complete list of the HSA series specifications.

Conclusion

An efficient Li-ion material must have characteristics of high charge/discharge ability, stable life cycle, and low cost for manufacturing. Furnace firing parameters are important in controlling the atmospheric conditions in order to protect from contamination at high temperatures. Reactions between the precursors need to be regulated with temperatures not exceeding the point where powder particles begin to increase in particle size and shape. Carbon coating must also be used in order to obtain good electronic conductivity within the system. It is often difficult to reproduce these methods with all the desired electrochemical performance necessary. Solid state reaction methods require long processing times with high temperatures which could cost in efficiency. Solution-based methods can be used instead with better results but with added costs and extra complexity. Furnace heat treatment can well be used for large-scale manufacturing procedures to obtain optimum lithium iron phosphate products.

References

1. Ren, J., Pu, W., He, X., Jiang, C., & Wan, C. (2011). A carbon-LiFePO4 nanocomposite as high-performance cathode material for lithium-ion batteries. *Ionics, 17*, 581–586.
2. Wang, Y., Liu, Z., & Zhou, S. (2011). An effective method for preparing uniform carbon coated nano-sized LiFePO4 particles. *Electrochimica Acta, 58*, 359–363.
3. Toprakci, O., Toprakci, H., Ji, L., & Zhang, X. (2010). Fabrication and electrochemical characteristics of LiFePO4 powders for lithium ion batteries. *KONA Powder and Particle Journal, 28*(28), 5–73.
4. Chang, Z.R., Li, H.J., Yuan, X.Z., & Wang, H.J. (2009). Synthesis and characterization of high density $LiFePO_2$/C composites as cathode materials for lithium-ion batteries. *Electrochimica Acta, 54*, 4595–4599.
5. Yamada, A., Chung, S., & Hinokuma, K. (2001). Optimized LiFePO4for lithium battery cathodes. *Journal of the Electrochemical Society, 148*, A224–A229.
6. Jugovic, D., Mitric, M., Kuzmanovic, M., Cvjeticanin, S. S., Cekic, B., Ivanovski, V., & Uskovic, D. (2011). Preparation of LiFePO4/C composites by co-precipitation in molten stearic acid. *Journal of Power Sources, 196*, 4613–4618.
7. Negamine, K., Reinsch, S., Mueller, R., Honma, T., & Komatsu, T. (2011). Crystallization behavior of lithium iron phosphate glass powders in different atmospheres. *Journal American Ceramic Society, 94*(9), 2890–2895.

Direct Bond Copper (DBC) Technologies

Ron Visser and John B. Snook

Introduction

Direct bond copper technology that primarily uses copper foil bonded directly to ceramic substrates was originally developed and patented by General Electric Company in the early 1970s and subsequently protected with a series of other patents. Although jewelry makers in fabrication of copper costume jewelry for many years have practiced a similar process, GE was able to patent the technology for its application to ceramics. The technology, which required royalty licensing, offers advantages over traditional thick- and thin-film metallization technologies for a wide range of modern microelectronic applications. The inherent advantages of using bulk metal copper foil in place of fritted or reactively bonded pastes and economically expensive sputtered metallization are increasingly finding wide-ranging applications across the span of military, commercial, industrial, and automotive industries as engineers readily accept the electrical and thermal properties of a traditionally favorite copper system for electronic circuits. The standard direct bond copper or DBC technology which uses copper foil thicknesses in the range of 0.005–0.020 in. is usually specified for high-power and high thermal management circuits where the large geometry requirements of 0.015 in.-wide lines and spacings can be used. New processes and additional refinements of the process which adapt the inherent advantages of the strong bonding mechanism of the DBC process, however, will allow the technology to be adapted for fine-line circuitry and will extend its applications to high-frequency circuits.

J. Xu et al., *Conveyor Belt Furnace Thermal Processing*,
https://doi.org/10.1007/978-3-319-69730-7_16

DBC Process

The DBC process uses OFHC (oxygen-free) or ETP (tough-pitch) copper foil which has been oxidized by chemical or thermal means to provide a uniform dark blue-gray Cu-O cupric oxide coating thickness, on the copper surface. This foil is placed on the face of an oxide ceramic such as alumina or beryllium oxide and introduced through an inert tunnel furnace, which is usually continually purged with nitrogen. From the phase diagram reproduced from an early GE patent 3,994,430 (Fig. 1), it is evident that a liquidus region at 1.54 atomic percent oxygen exists at 1065 °C. The belt furnace is profiled in a high-peaked bell shape thermal distribution to bring the copper and ceramic to a temperature above the 1065 °C eutectic bonding temperature at which the copper oxide forms a liquidus cuprous oxide (Cu_2O) matrix which wets and chemically bonds to the oxide surface of the ceramic. Since the eutectic temperature is 18 °C below the 1084 °C, melting point of pure copper, the furnace profile must be tightly controlled to overcome loading effects from both the mass of the entering substrates and the cooling effects of the nitrogen gases. The resultant bond is highly tenacious and rugged. With alumina and beryllium oxide surfaces, bond strengths exceeding 24,000 PSI are readily achieved. The conversion of the initial cupric oxide (Cu-O) formed on the copper and the resulting cuprous oxide (Cu_2O)-copper bond releases excess gas during the process which must be dissipated from the copper-ceramic interface to avoid gaseous entrapments under the copper which appear as blisters. These blisters, which are normally referred to as *tents* to distinguish them from reliability problems associated with thin-film blistering and plating blistering, are a commonplace phenomenon and are one of the highest yield point losses in the process. This problem is also compounded as the area of the bonded surface increases to the large alumina Herman (3.75×4.5″) and super HermanO (4.5×4.5″) sizes. To overcome this problem, the majority of DBC processes use copper foil that has been pre-etched on one side with a pattern of evenly spaced parallel grooves which when placed against the ceramic allow the dissipation of the gas during the eutectic formation preventing gas entrapment until the grooves become filled with eutectic themselves. Different vendors vary groove patterns and depths to optimize this technique for their oxidization process. These grooves although solving one problem lead to other subsequent process problems such as chemical entrapment during chemical etching, cleaning, or soldering processes, ceramic staining due to chemical entrapment during soldering or heat baking, and severe undercutting for fine-line circuit patterns. They can also become a source for moisture entrapment during environmental screening which can lead to a rupture of the copper to ceramic bond during temperature cycling and thermal shock conditions. More modern techniques allow control of the oxide thickness to eliminate the grooves from the copper foil completely to totally eliminate these problems. After the bond is formed, the copper and ceramic is slowly cooled in the furnace to temper the copper foil to a dead-soft annealed condition and to control the copper grain formation. Excess oxygen and high temperatures can lead to a roughened alligator-type skin condition due to high eutectic oxide segregating at the

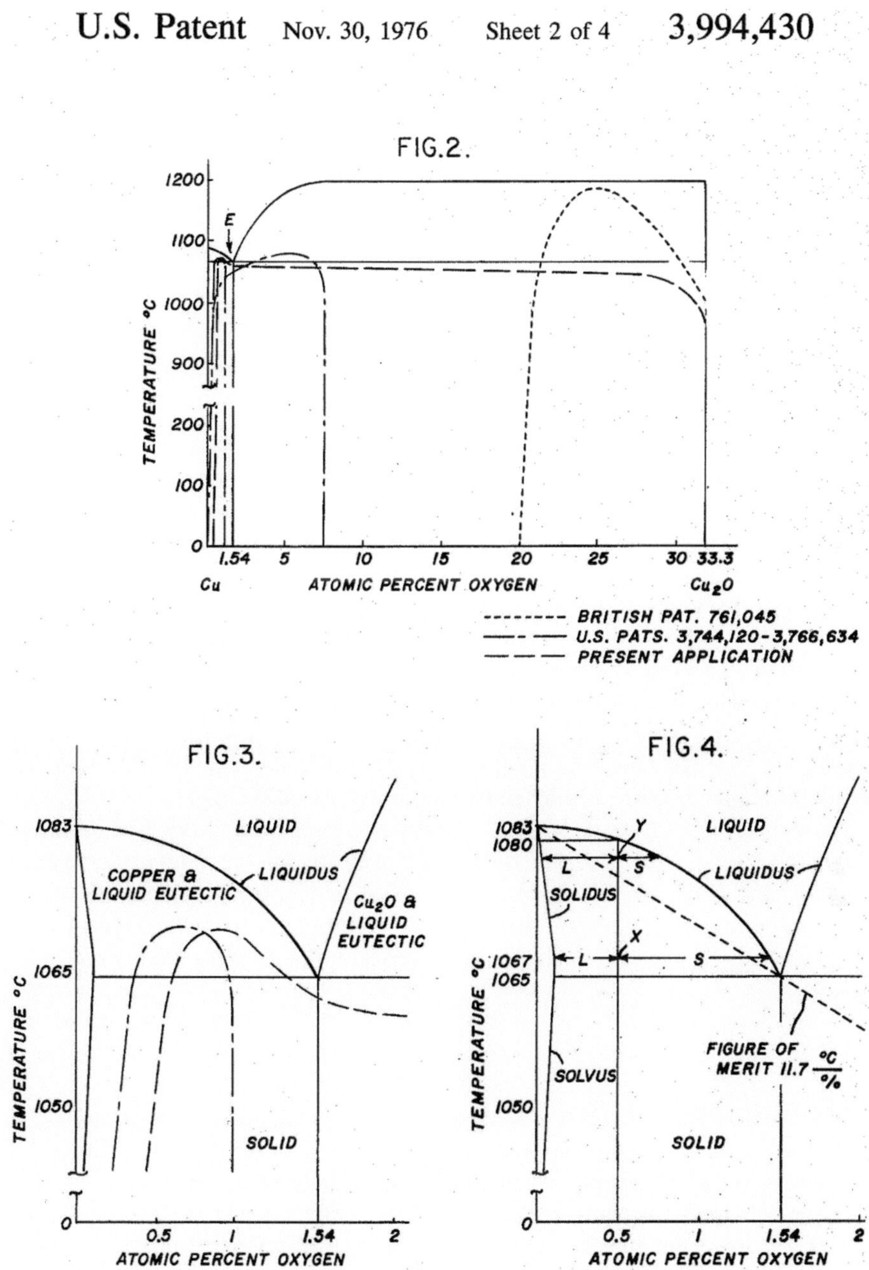

Fig. 1 Phase diagrams reproduced from an early GE patent 3,994,430 for Direct Bond Copper

Table 1 Thermal expansion and flexural strengths

Material	Thermal expansion coefficient	Flexural strength
Alumina (96%)	$6.3–7.4 \times 10^{-6}/°C$	420 mPa
Alumina (99.5%)	$6.7–7.1 \times 10^{-6}/°C$	360 mPa
Beryllium oxide (99.5%)	$8.47 \times 10^{-6}/°C$	230 mPa
Beryllium oxide (zirconia doped)	$7.6 \times 10^{-6}/°C$	350 mPa
Aluminum nitride	$4.6 \times 10^{-6}/°C$	350 mPa
Copper (99.9%)	$17.8 \times 10^{-6}/°C$	
Gallium arsenide		
Silicon	$4.0 \times 10^{-6}/°C$	
Kovar	$21.4 \times 10^{-6}/°C$	
Cu-Sil braze	$18.5 \times 10^{-6}/°C$	
In-Cu-Sil braze	$18.2 \times 10^{-6}/°C$	

copper grain boundaries. This condition can result in an uneven surface texture and extensive nodule formation, which may disturb chip attachment and wire bonding. The annealing of the copper during its cool down phase is important to overcome the severe thermal expansion mismatches of the copper and ceramic. Once bonded at DBC temperatures above 1065 °C, the large temperature differential to room temperature can lead to severe bowing and camber of the ceramic as the copper contracts at a faster rate. Proper selection of copper foil thickness to ceramic thickness with single- or double-sided bonding is important to match the thermal expansion coefficients and flexural strengths of the materials shown in Table 1. Since there is an effective zero yield point for ceramic materials, improper selection can result in ceramic fracture during temperature and thermal cycling.

To review, the DBC process yields exceptionally high strength bonds in excess of 24,000 PSI using copper foil with the bulk electrical and thermal properties of bulk copper and is bonded at an extremely high temperature, which permits substrates to be brought easily to Cu-Si and other high brazing temperatures without impairment to the bond. The rugged bond allows extensive rework to solder joints and wire bonds.

Patterned Substrates

The bulk of the DBC substrates manufactured are small single-chip versions used to mount power transistors and rectifiers and are typically less than 0.5″ square. These substrates are economically manufactured using the price advantage of small pressed BeO substrates in a one-up configuration where operators manually line pre-etched squares of copper on the ceramic substrates which are placed on carrier' bricks and sent through the bonding furnace. High-volume applications allow automated pick and place machines to eliminate the tedium and misalignment problems associated with operator assembly. More sophisticated geometries require chemically milled copper that is either pre-etched and then bonded to the substrate or is subsequently chemically milled after a full sheet of copper is bonded. The first approach requires copper bridges to hold the isolated islands of copper together for the bonding process, which must later be removed, and can also lead to distortions due to the large expansion rate of the copper. This approach also allows bonding of extended metallization for flying leads off the ceramic surface. The latter approach allows devices to be manufactured as multiple devices and a single substrate using batch-orientated printed circuit board techniques for photopatterning. Typically, dry film photoresist is applied to both sides of a copper clad substrate, and standard photolithography processes are used to develop the circuit pattern into the photoresist. Certain design rules are required to be incorporated into the emulsion masks used in the process to allow for proper etch factors dependent on the thickness of the copper foil used and to allow space for a dicing grid to permit later separation of the individual substrates with either a diamond dicing saw or laser techniques. Front-to-back alignment masks in a clamshell orientation are required for double-sided boards to permit a free copper area necessary for the diamond saw and laser techniques. Standard AutoCAD Gerber and DXF magnetic files are readily accommodated and are usually modified at the factory to add the details of etch factor, dicing grid, and alignment reticles. Twenty-four hour bulletin board modem service is frequently provided to permit rapid development of prototype circuits.

Plating

Although DBC substrates can be used directly in microelectronic applications without plating using aluminum wire bonding directly to the copper foil, many applications require nickel or nickel and gold plating. Copper foil which comes directly from the DBC furnace is typically passivated with a thin oxide coating protecting itself from rapid deterioration by room temperature oxidation. Parts are frequently shipped in this state and are protected in packaging with moisture desiccants to prevent excessive oxidation. Nitrogen storage or storage in desiccant cabinets is usually recommended to further protect the circuits. Surface preparation used in photolithography can reduce this surface protection, and parts are often protected with organic surface treatments. Plating of copper with nickel and gold can readily be achieved with electrolytic plating techniques where isolated island of copper areas can be electrically connected. For complex circuits with many isolated islands, specialized electroless nickel and gold plating techniques are adopted. Typical thicknesses specified are 50–150 μin of nickel and 10 μin nominal of gold flash. Specialized and proprietary processes are required for gold thickness that exceeds 10 μin.

Electrical Properties

The electrical properties of the DBC substrate assume the bulk characteristics of copper. Electrical resistivity of 2.5 μΩ·cm permits high-current loading. Twelve mil copper foil with 0.040″ line widths is capable of handling 100 amp continuous dc operations with low-temperature increases of less than 20 °C. Another important advantage of the high conductance of the copper film is its reduction in resistive or IR loading terms in the Vce(sat) conditions of power transistors. This reduction on Vce(sat) conditions can allow increased drive currents and reduce transient switching times.

Thermal Properties

The addition of bulk copper foil to both the top and bottom of the ceramic can rapidly reduce the thermal resistance and thereby reduce the temperature rise of the junction temperature of the semiconductor chip. Users have reported doubling the thermal conductance as well as electrical conductance by substitution of DBC heat spreaders in place of thick-film and refractory metalized ceramics. The reduction of junction operating temperature for semiconductors has a dramatic improvement in operating reliability and increases MTBF in Mil-Hdbk-217B calculations. This reduction also allows the designer to increase circuit densities and miniaturize the circuit elements. The copper foil has a dramatic effect on the effective thermal conductivity of the combination foil and ceramic. The copper foil assists in laterally spreading the heat along the surface of the substrate as well as increasing the spreading angle through the substrate. From a conduction model reported in ISHM monograph 6984-003 entitled *Circuit Board Material/Construction and its Effect on Thermal Conductivity* by Mr. Guy W. Wagner, the effective conductivity of a composite substrate can be calculated by the equation:

$$K_e = \frac{K_1 t_1 + K_2 t_2 + \ldots + K_n t_n}{t_1 + t_2 + \ldots + t_n}$$

where

K = thermal conductivity of each material
t = thickness of each material

The thermal conductivities of common microelectronic materials are listed in Table 2. The effective thermal conductivity of cladding copper foil of differing thicknesses onto differing ceramic thicknesses is calculated in Table 3. As is evident in this table, large improvements can be made by the excellent thermal properties of the copper foil. Whereas the ratio of thermal conductivities of alumina to beryllia is

Table 2 Thermal conductivities

Material	Thermal conductivity (W/mK)		
	25 °C	150 °C	200 °C
Alumina (96%)	20	15′	
Alumina (99.5%)	34	25	
Beryllium oxide (99.5%)	260	181	151
Beryllium oxide (zirconia doped)	240	162	139
Copper	386		
Gold	314		
Silicon	125		
Kovar	17		
Solder (60/40)	49		
Air	0.03		

Table 3 Thermal resistance calculations

Material	Thickness	Copper thickness	Equivalent thermal conductivity (W/mK)	
			One-side copper	Two-side copper
96% alumina	0.010	0.005	142	203
	0.010	0.008	183	245
	0.010	0.010	203	264
	0.010	0.012	220	278
	0.015	0.005	112	166
	0.015	0.008	147	209
	0.015	0.010	166	229
	0.015	0.012	183	245
	0.025	0.005	81	125
	0.025	0.008	109	163
	0.025	0.010	125	183
	0.025	0.012	139	199
99.5% alumina	0.010	0.005	151	210
	0.010	0.008	190	251
	0.010	0.010	210	269
	0.010	0.012	226	282
	0.015	0.005	122	175
	0.015	0.008	156	216
	0.015	0.010	175	235
	0.015	0.012	190	251
	0.025	0.005	93	135
	0.025	0.008	119	171
	0.025	0.010	135	190
	0.025	0.012	148	206

over ten times for the raw material, the influence of the copper cladding process will narrow this gap near a two to one ratio. Many applications can substitute economical alumina for the more expensive BeO. Designers who have learned to economize by minimizing metal areas in thick- and thin-film circuits to reduce, precious metal costs can increase thermal performance by only removing copper foil where electrically necessary. This technique gives DBC its distinctive appearance in circuit designs with its massive areas over the sparsely designed traditional circuitry.

Another effect associated with the copper cladding is its low transient thermal response. Due to mass and high thermal capacity of the copper foil, semiconductors can be run at high peak pulses and short surge currents with a slow response in thermal heating effects.

Mechanical Properties

The tenacious bonding of the copper foil to ceramic allows repeated rework to soldered connections and to wire bonding applications. Unlike the fragility of thick- and thin-film circuits, the DBC circuits allow rough handling and can endure repeated eutectic bonding temperatures. The selection of the ratio of copper thickness to ceramic thickness is an important consideration in the initial design of the circuit. One-sided bonding on thin substrates can cause severe camber conditions that can degrade the bond with repeated temperature cycling due to shear and *bimetallic*-type bending forces or lead to ceramic fracturing. These effects can be reduced by double-sided bonding to equalize the forces between opposite faces of the ceramic. Leaving copper in nonelectrical areas will not only improve lateral heat spreading but also minimize bending stresses during temperature cycling. Since excessive camber can lead to solder voiding during substrate attachment and subsequent degradation of thermal performance due to air entrapment, the ceramic is often designed to be thicker than a traditional thick- or thin-film circuit. With the improved thermal performance of the copper, this increase in dielectric thickness can minimize cambering leading to improve substrate attachment and less bending stresses on the copper foil.

Future Development

The exceptional bonding strength and high reliability of the DBC process combined with the electrical and thermal performance of copper conductor circuitry have taken DBC from early prototypes to production quantities for high-power, high-current, and high-temperature applications. New DBC-related processes are continually being developed to meet the challenges of other technologies. A few of these applications include:

1. Power Packaging

Special hermetic copper power feedthrough techniques are being tested to permit high-current applications with low IR conductance losses. Power feedthroughs capable of handling over 100 amps with less than 100 μΩ resistances are being designed.

2. Power Resistors

The DBC techniques are not limited to copper oxide eutectics but can work with other metallic oxides. Special bulk alloy foils used as resistor elements are being eutectic alloy bonded directly to ceramics to take advantage of the exceptional power handling capabilities of the process and to incorporate power RF terminations.

3. Microwave

The thickness of the copper foil however limits fine-line geometries and repeatable etching resolutions for high-frequency applications and large digital circuitry. New developments in prototype development permit a combination of DBC and pattern plate-up techniques using printed circuit board technologies of acid copper plating to offer products to span the microwave and digital fields.

4. Non-oxide Ceramic Bonding

Special processes are required to bond other non-oxide ceramics. Aluminum nitride which offers an attractive alternative to the inherent economic and environmental disadvantages of beryllium oxide requires specialized surface oxidation to permit DBC bonding. Most thermally grown oxides on AIN result in a bauxite alumina coating that is porous and weakly adherent to the underlying AIN surface. Not only is this interface a poor thermal conductor between the copper and AIN materials, but it is also mismatched in thermal expansion characteristics between the materials. The high stresses induced by the mismatch with copper in the high bonding temperatures of the process as well as the reduction effects' of the inert atmosphere of the nitrogen gas used in the process result in a stressed interface. Repeated temperature cycling and thermal shock can easily rupture this interface with loss of adhesion occurring at the alumina to aluminum nitride interface. Specialized techniques to achieve a graded and highly dense but thin interface are required to optimize the bonding between the AIN and its oxide surface.

Conclusion

Direct bond copper (DBC) technology from its earlier applications in appliance modules and refrigeration controls finds increasing usage in high-technology hybrid power applications. From advanced radar systems and submarine sonar circuitry to industrial power control and automotive applications, DBC is capable of handling the extremely high electrical current and high thermal cooling requirements that cannot be done with conventional thin- and thick-film metallization technologies. DBC has developed alongside with recent advances in power semiconductors that eliminate the old power tube devices to supply the hybrid designer with a cost-effective microelectronic and packaging technology. Applications for high current, high voltage, high power, and high temperatures can all be accommodated by a low-cost, reworkable metalized substrate material that will become the standard in the next generation of electronics.

Influence of Firing Temperature and Atmospheric Conditions on Processing of Direct Bond Copper (DBC)

Introduction

The development of DBC has pushed boundaries of power electronic applications. It is used in such products as concentrated photovoltaics and semiconductor modules for automotive applications [1]. The value of DBC comes from its high thermal conductivity due to copper and its low coefficient of thermal expansion as a result of ceramic. The need for low thermal expansion is so solder on silicon semiconductors which will not be damaged with heat cycling. For DBC ceramics, alumina (Al_2O_3) is commonly used, but other ceramics like aluminum nitride (AlN) or beryllium oxide (BeO) are also utilized in bonding. Alumina is preferred because of its low cost and efficiency compared to these other ceramics, and DBC is chosen over other bonding processes because there is no need for intermediary layers in the process. According to Burgess et al., Al., MgO, SiO_2, and CaO are intermediary layers located at the grain boundaries of the ceramic. This layer penetrates into the porous layer of the metal at sintering temperatures and locks the metal into the ceramic. DBC is preferred because intermediary layers often have problems with current flows at the metal ceramic interface and are less corrosion resistant leading to early failure.

DBC begins with layering of copper onto alumina as seen in Fig. 1. The contact interface between the copper and the alumina is the most important area of processing. Copper exhibits a wetting behavior on alumina that covers the area of contact during firing. With a temperature range between 1065 and 1083 °C, interface adhesion occurs. With research and developments in DBC technology, production is possible on a larger scale using equipment such as the belt furnace. Controlling furnace conditions is important in reaching the high eutectic temperature needed to begin the bonding process while staying below the melt temperature of copper. Specific controls of the furnace atmosphere are needed as well during processing.

J. Xu et al., *Conveyor Belt Furnace Thermal Processing*,
https://doi.org/10.1007/978-3-319-69730-7_17

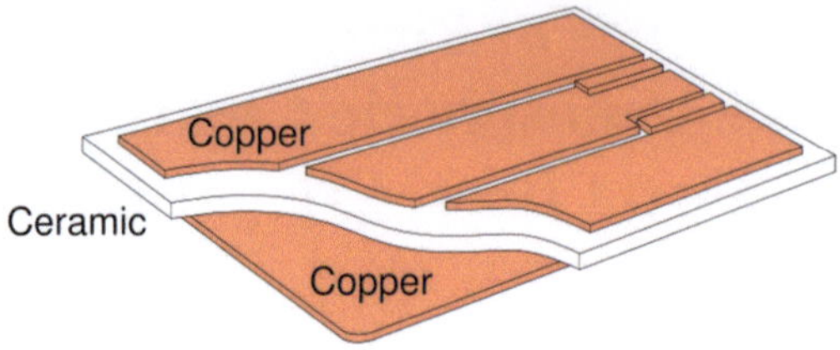

Fig. 1 Ceramic-copper bond representation [2]

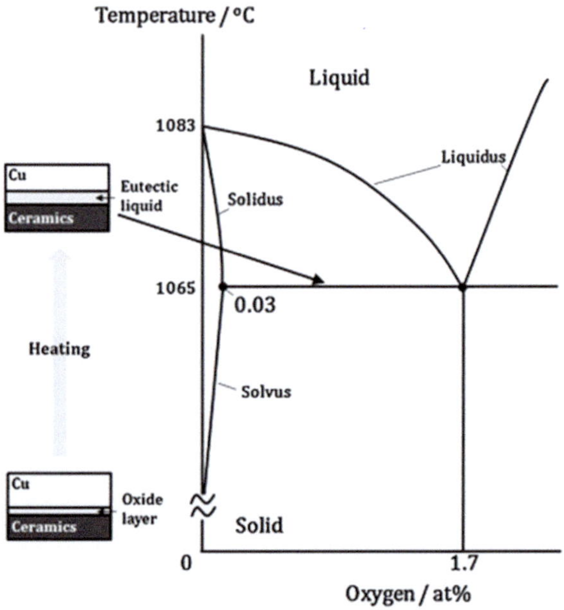

Fig. 2 Eutectic point and percent concentration of O_2 [3]

Process of DBC

The process begins with applying copper onto alumina inside an atmosphere controlled furnace. The ideal settings for the atmosphere are flowing inert gas, such as nitrogen or argon, and a small percentage of oxygen. The eutectic that occurs between copper and oxygen is found to be around 1065 °C as seen in Fig. 2.

According to Faghihi et al. [4], in their process of making DBC, the copper material used was 99.99% pure and alumina was at least 97% pure. Copper strips of 0.8 mm in thickness were observed. It will be discussed later on the important effect of copper thickness on peel strength. For cleaning, copper strips are degreased and washed using treatments such as detergent and hot water scrubbing. Alumina is cleaned by polishing, ultrasonic treatment in ethyl alcohol, and annealed in air at 1000 °C to reduce hydroxyl groups [4]. Oxidation takes place after cleaning which involves the heating of copper creating a thin film layer of cuprous oxide (Cu_2O) at the interface between the alumina and copper. This can be done prior to firing with the ceramic or when the copper is already laid onto the ceramic. Either way the Cu/Cu_2O layer serves as a bonding agent with the ceramic.

There are varying rates of heating the substrate as seen by Ning et al. [5], Faghihi et al. [4], or Burgess et al. [2]. In studies done by Faghihi et al., the first stage of the bonding process occurs with a total inert gas atmosphere, such as argon or nitrogen, with a small percentage of oxygen mixed in as defined by the eutectic. In this atmosphere, the substrate is heated to 1000 °C at 10 °C/min. Then at a slower rate of 2 °C/min, the substrate is heated to 1075 °C. At this point, the hold temperature of 1075 °C is kept constant for an hour. This is the time where the alumina copper bond is formed. The substrate is then cooled down to 400 °C and held at that temperature for an hour to reduce thermal stresses and finally cooled down to room temperature. This process requires proper control of temperature during the whole process. Likewise, atmospheric conditions need to be observed and controlled. A final stage of vacuum processing increases the pore size and decreases the peel strength resulting in a substrate that will be more susceptible to failure with high loads [4]. Another before-mentioned atmospheric condition needed is a 1.4 mole% of oxygen when heating the substrate up to the hold temperature of 1065–1075 °C. This peak temperature needs to be lower than 1083 °C, the melt temperature of copper [6]. During cooling is where the diffusion of oxygen out of copper is taken place. At this stage, an atmosphere of H_2 can be used for cooling. The hydrogen gas that is injected into the furnace helps remove the Cu_2O particles at the interface of the eutectic zone increasing bond strength due to the Cu/Al_2O_3 interlock. Another temperature cycling method for DBC is shown below in Fig. 3. Note in this method there is no hold time at 400 °C during cooling.

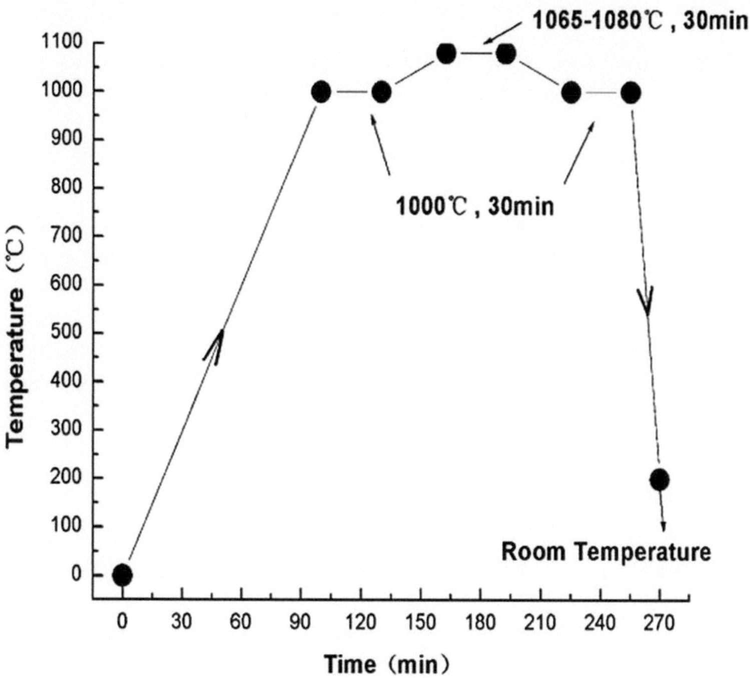

Fig. 3 Time-temperature of DBC substrate bonding [6]

Characteristic Features of DBC

Copper allows the feature of high thermal conductivity along with good characteristics in electrical conductance. The copper foil used in substrates range from thicknesses of 0.125–0.8 mm. This thickness is important for quality. Research shows thinner 0.3 mm thick copper sheets were used because of lower occurrences of crack propagation when compared with larger thicknesses [7]. With a 0.3 mm thick copper foil fused to alumina, the coefficient of thermal expansion is only 7.2×10^{-6}, slightly higher than ceramic by itself at 6.8×10^{-6} [3]. Another noted benefit is a better hermetically sealed package for power modules. With DBC, a power module package can be as much as 70% less in weight for industry, and new developments such as fluid cooled substrates are able to dissipate heat up to 400 W/cm^2. A power module is represented in the following figure. T_a and T_b denote the temperatures on either side of the module at the copper (Fig. 4).

Effect of Temperature and Atmosphere in Bonding

Oxidation of the copper is one of the most important factors of bonding the copper to alumina. The interface needs a thin film layer of Cu-Cu_2O, achieved by firing to the eutectic temperature and creating favorable adhesion to alumina. According to studies done by Ning et al. [5], the oxidation temperature that will create a good oxide layer is 300 °C for 60 min. Wetting of the Cu-Cu_2O onto the alumina is important for adhesion strength. This proper wetting of alumina ensures that the Cu-Al_2O_3 bond is established. During firing of the alumina and copper substrate, there needs to be an atmosphere with a small amount of oxygen. The oxygen content is 0.39% by weight (also mentioned before as 1.4 mol% O_2) as seen on the Cu-O binary phase diagram of Fig. 5. At this percentage of oxygen, the eutectic occurs at 1065 °C.

The effect of H_2 is important during cooling but can have negative consequences. Hydrogen atmosphere can weaken interface peel strength with temperatures above 400 °C. But at lower temperatures, H_2 atmosphere can help reduce the cuprous oxide (Cu-Cu_2O) increasing the strength of the substrate. The peel strength of specimens studied show how peel strength is raised from 13.1 to 17.1 kg/cm due this stage.

Firing Operations Using a Belt Furnace

The HSA series furnace is designed to meet the requirements needed to produce DBC. This furnace is able to reach maximum temperatures of 1150 °C. The HSA series furnace comes with a refractory heating chamber equipped with ceramic fiber FEC (fully enclosed coil) heating board. The heating works to give fast thermal

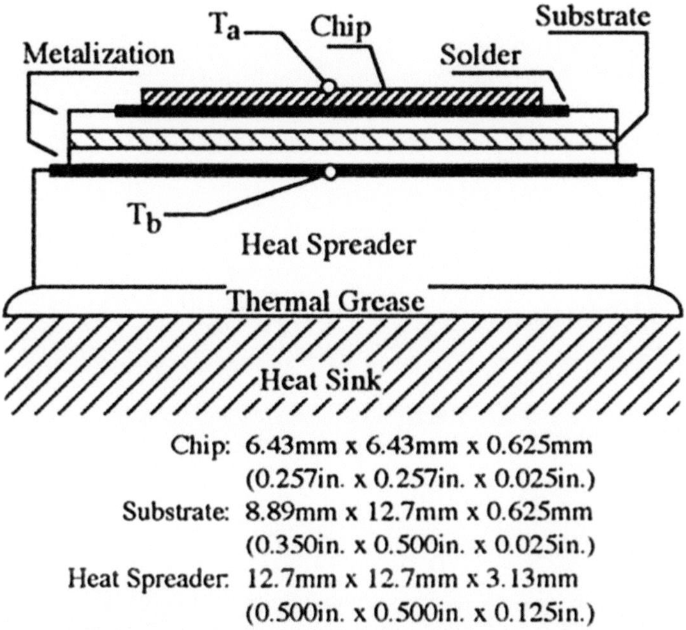

Fig. 4 Generic structure showing substrate in relation to other components [7]

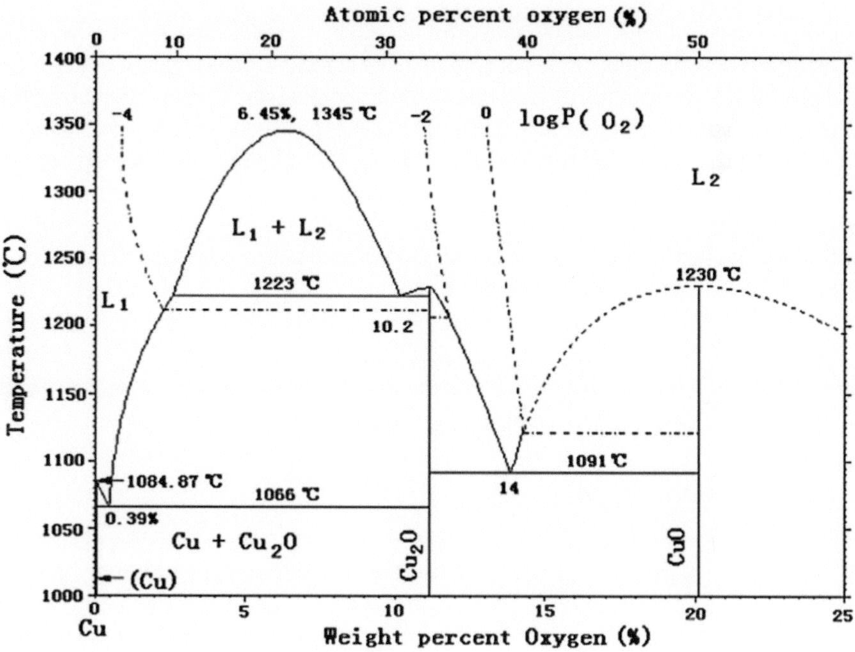

Fig. 5 Cu-O binary phase diagram [5]

response. It is able to work in the temperature range of 1065–1083 °C. Forced air or water cooling is used in the cooling section of the furnace. The muffle design located within the furnace helps with control of atmospheric conditions and also helps in maintaining a cleaner environment inside the furnace. Belt cleaning is done through use of a steel brush to clean the belt. Ultrasound belt cleaning is also available for cleaning needs as an extra option.

A microprocessor-based PID controller provides appropriate system control. Type K thermocouples are used in determining the zone temperatures, and the controls are located on the right-hand side of the furnace which can be viewed from the entrance. The central processing unit (CPU) is located under the exit table and is available with a windows operating system for ease of use. A program is installed ready to control furnace parameters such as belt speed, zone temperatures, and atmospheric conditions. Temperature profiles can be stored and retrieved as well for future purposes. Thermocouple ports are located at the entrance table for connecting the profiling thermocouple directly into the microprocessor. This feature allows for monitoring and recording of actual temperatures experienced by the part. Included in the software are programs for capturing/storing, displaying, and printing out the furnace profile. Additionally, the furnace is equipped with a redundant overheat safety protection system which incorporates an additional type "K" thermocouple in the center of each controlled zone and the multi-loop alarm. The specifications of a HSA 2006-0812ZNO is given below.

Specification	HSA2006-0812ZNO
Rate temperature	1180 °C Normal operating temp: RT-1100 °C
Belt width	200 mm/8″
Above belt clearance	60 mm/2.4″
Heating length	3600 mm/142″
Insulation cooling length	400 mm/15.7″
Water cooling length	2200 mm/86.6″
Air cooling length	200 mm/8″
Control zones	8
Conveyor speed	25–150 mm(1–6″)/min

Specification	HSA2006-0812ZNO
Heating elements	Ceramic FEC heating board
Insulation material	High quality ceramic fiber
Overall system width	1200 mm/47″
Overall system length	9350 mm/23 ft
Overall system height	1350 mm/53″
Typical temp. uniformity	±2 °C
Controlled atmosphere	Nitrogen is standard, oxygen as required
Power	480 V, 3 phase, 5 wire, 60 Hz, 48 kW Heat insulation power draw: 18 kW
Spare parts	1 set of cooling fans, 1 solid state relay, 2 sets of heating elements

Conclusion

Traditional substrates use buffer layers such as molybdenum to lower the thermal expansion coefficient. DBC has taken the place of this giving lower cost and higher production. The proposed HSA furnace has the capabilities to elevate to higher temperatures, maintain cross-belt uniformity, and achieve the required atmospheric conditions. At the same time, the muffle design allows the flowing of specific gasses into the required zones. To conclude, the HSA is an ideal furnace for production of quality DBC products.

References

1. Curamik. *Curamik product information*. [Brochure].
2. Buttay, C. *Cross section of DBC*. [Image].
3. Schulz-Harder, J., & Exel, K. (2003). *Recent developments of direct bonded copper (DBC) substrates for power modules*. In Fifth International Conference on Electronic Packaging Technology Proceedings. Shanghai: Curamik Electronics GMBH.
4. Faghihi, M. A., Ghasemi, H., Kokabi, A. H., & Riazi, Z. (2009). Alumina-copper eutectic bond strength: Contribution of preoxidation, cuprous oxides particles and pores. *Transaction B: Mechanical Engineering, 16*(3), 263–268.
5. Ning, H., Ma, J., Huang, F., Wang, Y., Li, Q., & Li, X. (2003). *Preoxidation of the Cu layer in direct bonding technology*. Beijing: Department of Materials Science Engineering, Tsinghua University.
6. Hopkins, D., Bhavnani, S., & Dalal, K. (1992). Thermal performance comparison and metallurgy of direct copper bonded AlN, Al2O3, and BeO Assemblies. In *ISHM 1992 proceedings*.
7. Schulz-Harder, J., & Exel, K. (2006) Advanced DBC (direct bonded copper) substrates for high power and high voltage electronics. In *22nd IEEE SEMI-THERM symposium*. Curamik Electronics GMBH.

Ultrasonic Belt Cleaning

Introduction

Ultrasonic belt cleaners are labor-saving appliances, which collect particles that come off of the furnace belt so that these contaminants can be safely drained away. Ultrasonic cleaners combine ultrasound technology with a sanitizing solvent, which thoroughly scours the furnace belt to remove all traces of contamination. Ultrasonic technology is very important for furnace belt maintenance, as it can deliver incredibly precise and efficient cleaning capabilities. Preserving the cleanliness of a furnace belt is vital when it comes to prolonging the lifespan of a furnace, and ultrasonic cleaning has proven to be the best solution to long-term furnace belt care.

Principles Behind Ultrasonic Cleaning Technology

An ultrasonic belt cleaner uses carefully positioned transducers to emit sound waves into the cleaning tank, which contains a purifying solvent. A transducer is an instrument that transforms one form of energy into another. In this case, transducers produce ultrasonic waves that cause the solvent to vibrate at ultrasonic frequencies, generating bubbles that implode around the sections of the belt in need of sanitation.

These bubbles are specifically formed by extremely rapid pressure changes within the fluid, caused by the ultrasonic waves. These waves travel through the tank, causing the liquid to compress and expand. As a result, an immense amount of microscopic bubbles is left behind, which implode and dislodge impurities from the furnace belt (Fig. 1).

J. Xu et al., *Conveyor Belt Furnace Thermal Processing*,
https://doi.org/10.1007/978-3-319-69730-7_18

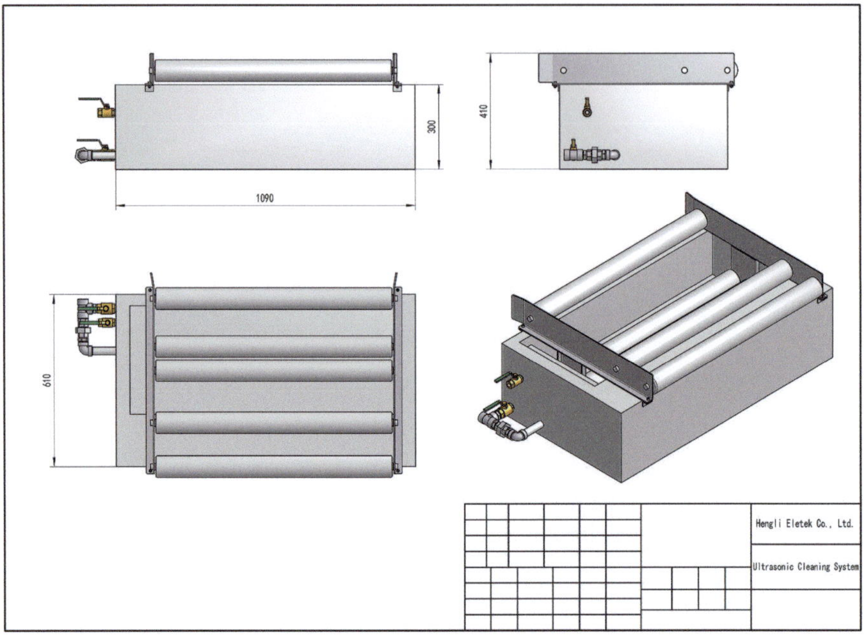

Fig. 1 Ultrasonic belt cleaner schematic

Process of Cavitation: A Microscopic Brush

Water can only exist in a liquid state above a minimum pressure (the threshold of cavitation). When pressure decreases below this threshold, water becomes a gas and bubbles form. Then, when the pressure increases, gas bubbles collapse again. This process of creating rapid formations and destructive prostrations of minute bubbles in liquid is known as cavitation, and it is responsible for the scrubbing action that removes contaminants from the belt. Cavitation will transpire throughout the cleaning solvent if the energy intensity is adequate. It will also expedite chemical reactions and the rate at which surface films are dissolved.

As the ultrasonic frequencies increase, the number of microscopic bubbles also increases; however, the respective energy that is released from each bubble simultaneously decreases. This means that higher frequencies are ideal for small particle removal, without damaging the belt. The vast amount of imploding bubbles creates a highly effective scrubbing action on both the exposed and the hidden areas of the belt that are immersed within the cleaning solution.

Advantages of Ultrasonic Belt Cleaning

Simply put, the advantages of using ultrasonic belt cleaning technology revolve around precision, speed, and consistency. Ultrasonic energy is able to infiltrate obscure regions of the belt, which means that all areas of the belt will be comprehensively cleansed. Furthermore, integrating the belt cleaner into the furnace is straight-forward. The ultrasonic cleaning is in-line, so the belt travels directly through the tank, and there is no need to disassemble the belt from the furnace. The belt can pass through the tank whenever necessary. This means superior cleanliness and will guarantee a better working environment for the furnace.

Ultrasonic cleaning also works faster than any other conventional cleaning procedure in the elimination of contaminants. Due to the efficiency of ultrasonic antisepsis, labor-saving and sanitization benefits indicate that ultrasonic technology is the best economical method for cleaning furnace belts. When ultrasonic technology is used, there is far less risk of products getting contaminated, which improves yield. Furthermore, there is less belt abrasion, which means that the belt will have a longer lifespan. Lastly, there is no need to ever replace any brushes.

Ultrasonic technology offers incomparable cleaning consistency throughout the entire belt. Using a steel brush as a tool for cleaning the belt, on the other hand, can often be time-consuming, imprecise, and inconsistent. A steel brush uses steel wire bristles to clean the surface area around the belt, and due to its abrasive nature and inability to reach abstruse regions of the belt, it cannot reach the level of cleanliness that an ultrasonic belt cleaner can easily achieve.

Equipment Maintenance and Belt Cleaning Guidelines

- Keep the belt speed under 150 mm/min while cleaning.
- Use distilled water instead of tap water as the solvent. Tap water can contain impurities, and distilled water that has been degassed allows for more even distribution of cavitation.
- Always monitor the quality of the solvent.
- After use, clean out the rinse tanks.
- After use, change the filters.

How Challenging Conventional Wisdom Can Optimize Solder Reflow

Marc Peo

With solder reflow, relying on commonly accepted practices can lead engineers to overlook some relatively simple factors that are critical to optimizing the process. An examination of six "conventional wisdom" tenets reveals that challenging them can contribute significantly to improved yields.

A Standard Profile?

Figure 1 includes the "ideal" reflow profile. It is characterized by a 2–4 °C/s ramp to a dwell zone of 30–90 s at approximately 150 °C, followed by a second ramp to about 210 °C. This graph does represent a satisfactory profile for certain pastes and some ovens, but it is by no means suitable for all situations. Applications for which the (a) profiles are appropriate include pastes that require a dwell time at 150–160 °C to allow for flux activation and for IR-based ovens that create large temperature differentials (ΔT) on the surface of the PCB. In the latter case, the time in the dwell zone reduces the ΔT by permitting lower temperature areas to "catch up" to the higher thermal areas, at which point the entire board achieves near equilibrium.

Today, however, many no-clean pastes are used with greater frequency and actually require a reflow profile without a dwell zone. The typical form for such a profile, Fig. 1b, is commonly referred to as a "tent" or "straight ramp" profile. For no-clean pastes, which have less active fluxes, excessive dwell time serves only to deplete the flux before reflow.

A second advantage of the tent profile is that it speeds the reflow process, since residence in the oven is reduced by the time formerly required for the dwell zone. Hence, when typical 4- to 5-min profiles can be completed in 3–4 min, throughput increases of 20–25 % are likely. On the other hand, a shorter oven also can be used to achieve the same throughput and to conserve factory floor space.

J. Xu et al., *Conveyor Belt Furnace Thermal Processing*,
https://doi.org/10.1007/978-3-319-69730-7_19

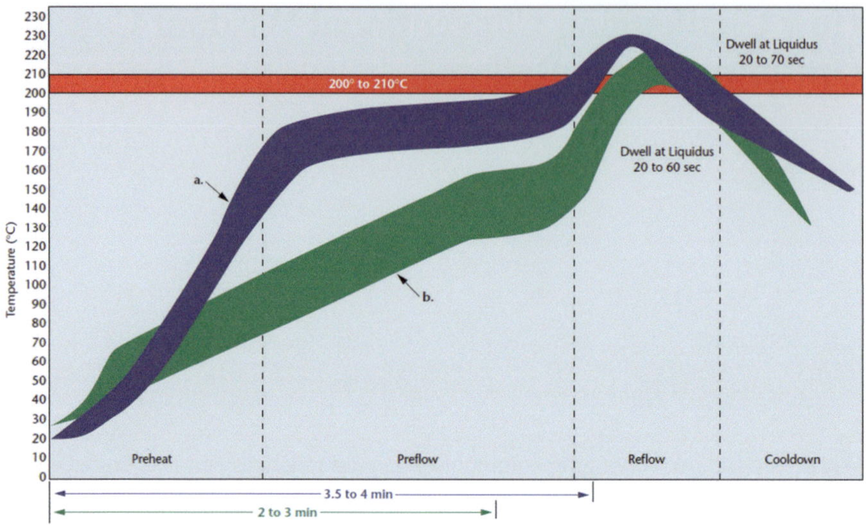

Fig. 1 The "standard" reflow profile (a) is compared to the "tent" or "straight ramp" pattern. While the former is suitable for solder pastes that require dwells at specific temperatures, the growth in the use of no-clean pastes demands more controlled high-temperature periods lest the flux be depleted

Avoid Reflow of Double-Sided PCBs?

Reflowing double-sided boards in a full convection oven is generally proscribed since it can cause components to be blown off the bottom. In fact, all full convection reflow ovens today are engineered to account for the special requirements of double-sided reflow. They include improved controls for interior air velocity, combined with the surface tension of solder paste, to hold bottom-side components in place and to eliminate the possibility of blowing the parts off during double-sided reflow.

Additionally, after eutectic solder is first melted, the alloying that occurs at the solder joint during reflow effectively increases the remelting point. Here, the alloy no longer has eutectic properties, making it uncommon for bottom-side components to be released during a second run through the oven. In some applications, components as large as 68-pin PLCCs can even be suspended on the bottom purely via surface tension of the solder. Thus, a formula for secondary side mounting can determine a component's candidacy for bottom-side attachment [2]:

$$C_g / P_a$$

where C = component's weight in grams and P = total pad area in square inches (g/in^2 must be $\leq$30 for second-side mount).

To anchor components larger than 68-pin PLCCs (e.g., relays and other high mass/low lead count packages), most PCB assemblers use an epoxy.

In other cases, solders with different melting points can be used. For example, if the bottom side is reflowed first using a high melt point solder (greater than 183 °C), then the top side can be reflowed using a eutectic solder (melting point, 183 °C). With this procedure, the bottom side cannot be reflowed a second time. Communicating such issues back to the design stage (and developing strategies to deal with them) is an important factor in resolving process challenges.

What Really Causes "Tombstoning"?

Tombstoning, or the displacement of a chip component during reflow into a near vertical position with only one terminal soldered, is caused by accelerated heating: when exceptionally high ramp rates (e.g., 6–7 °C/s) are in effect, the solvents in the paste can boil, creating bubbles with sufficient force to pop components out of place.

This was certainly a possibility until recently. However, the development of water-insoluble solvents having higher boiling points has effectively eliminated the bubbling issue. Nor do water-soluble solvents attract moisture that can boil off as heating progresses. Instead, tombstoning can be caused by the following conditions that can occur at various stages of the assembly process: inaccurate component placement. As shown in Fig. 2a, the greater wetting force exerted on one end of an incorrectly placed component will pull it up on the end, creating the characteristic "Manhattan" or tombstoning effect.

Excessive Pad Size. A greater amount of paste relative to the surface area of the part's terminal can also result in tombstoning (Fig. 2b). Also, a greater wetting force can be created during reflow in the case of an uneven deposition of paste on the pads.

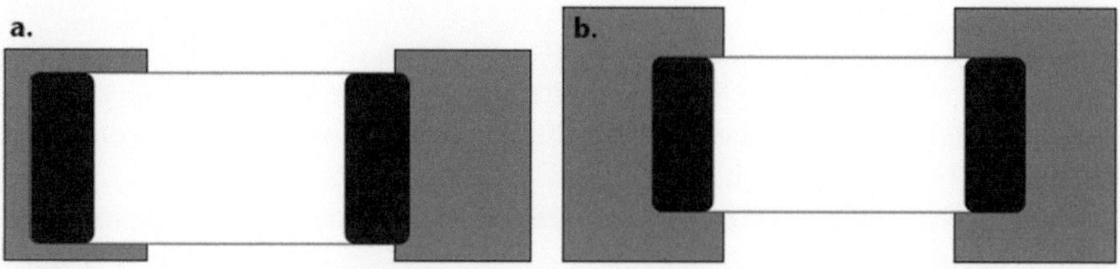

Fig. 2 Conditions that lend to the possibility of component tombstoning during reflow include inaccurate placement (**a**) in which a greater wetting force at one end serves to erect the opposite end and (**b**) excessive pad size vs. the part's terminal size, leading to the same effect

Using Nitrogen (Vs. Air) Improves the Process?

In fact, a nitrogen atmosphere in the oven does improve reflow results. It creates solder joints that are shinier, improves wetting angles, and increases the "margin for error" in the process. In addition, certain applications such as ultralow residue fluxes actually require nitrogen owing to their lower levels of activity. However, in applications where nitrogen is not stipulated, many assemblers have found the gas to be a "Band-Aid" in compensating for situations that can be fixed by careful process control.

Poor soldering can result from many causes including improper deposition of paste, use of an inappropriate paste, paste that has been on the stencil too long, inaccurate component placement, or an improper reflow profile. While nitrogen reflow may mask the effects of those problems, a simple process adjustment may be all that is required to improve results. More importantly, making process control adjustments carries no ongoing costs, while nitrogen usage is certainly expensive, i.e., there are usages, facilities, and setup charges to be accounted for with N2 processing. Accordingly, it is necessary to review costs vs. benefits before committing to the process.

During the review, nitrogen consumption becomes paramount. Obviously, reflow ovens that consume less nitrogen will cost less to operate, as shown in the table, and are easier to justify when they are required.

Lower Oxygen, Better Results?

Field testing has shown virtually no difference in quality from reflow units whose oxygen levels are held from 15 to 100 ppm. In fact, wetting angles and joint strengths have been found to be basically identical. In certain cases, it has even been reported that wetting forces at very low oxygen levels may actually be too strong, creating yet another cause of tombstoning on small parts. Since running at a higher oxygen level makes no difference to the overall process (while operation at the very lowest consumes significant amounts of nitrogen), this is certainly an area in which cost savings can be realized, as shown in Fig. 3.

Certain ovens offer a closed-loop nitrogen controller that permits a minimization of the gas' consumption by maintaining a specific ppm oxygen level. These systems regulate nitrogen flow automatically and compensate for changes in board load, ambient air conditions, and other factors by maintaining ppm levels within an acceptable tolerance band.

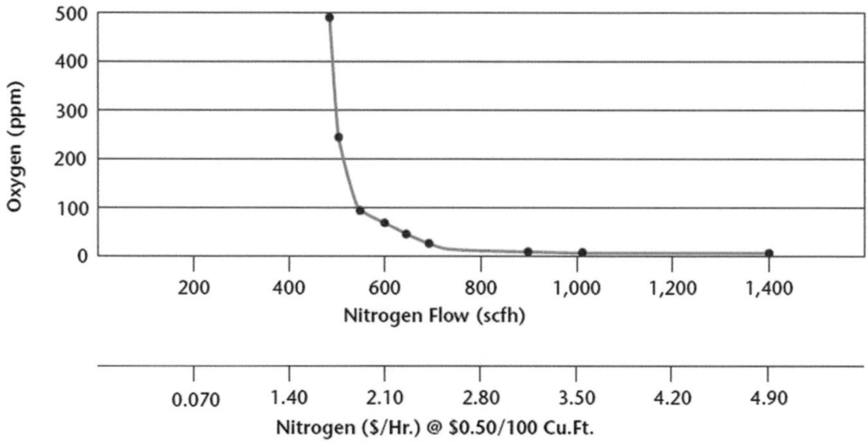

Fig. 3 How the cost in nitrogen consumption escalates as the oxygen ppm level is reduced in a reflow oven's atmosphere

Processing in N2 Means More Maintenance/Downtime?

With nitrogen processing, the cooling zones cause flux condensation, and it is thought, more frequent maintenance, and increased downtime. While the temperature of the water in the heat exchanger does cause flux condensation, several systems now address this issue: flux burn-off. Certain ovens provide a periodic burn-off system that heats automatically to vaporize flux as it accumulates. The flux turns into a fine ash and is exhausted with no ill effect on the process. Flux filtration: another system traps and removes flux before its entry into the heat exchanger. Result: very little flux enters the cooling chamber to condense on the heat exchanger. The flux filters may be removed and replaced even while the oven is running, eliminating the need for downtime to perform this maintenance. Waterless cooling eliminates recirculated water heat exchangers, thus removing the possibility of condensation inside the oven. In this enclosed system, the flux-laden gas is diverted from the heating zones and precipitated out using a cyclonic separation system and air-to-air heat exchanger. The cooled, flux-free gas is then reintroduced into the cooling zone as the flux is collected outside the oven chamber, again allowing preventive maintenance to be performed while the oven is running and eliminating the need for maintenance downtime.

A review of these commonly accepted reflow practices indicates that the surface mount process often defies generalization. In many cases, each process is unique and requires individually designed methods to satisfy its own requirements. And, of course, there is no substitute for rigorous process control. However, manufacturers are not without resources to assist in fine-tuning their processes. To take advantage of rapid developments in advanced technology, assemblers should work in partnership with all their vendors including suppliers of placement equipment, screen printers, solder paste, conveyors, and reflow ovens. The sales and technical support

Table 1 Worldwide cost of nitrogen in reflow vs. consumption

Country	Base cost/ft^3	AOC 2000 scfh[a]	AOC 600 scfh[a]
United States	$0.0035	$14,560	$3,640
Continental Europe	0.0070	29,120	7,280
United Kingdom	0.0070	29,120	7,280
Scandinavia	0.0018	7,280	1,820
Mexico	0.0070	29,120	7,280
Thailand	0.0105	43,680	10,920
Israel	0.0175	72,800	18,200
Singapore	0.0350	145,600	36,400

Cost's source: ITM

[a]Annual operating cost, continuous single shift, 2080 hr/yr

staffs of these companies have experienced literally thousands of applications. Hence, their expertise can provide valuable process contributions usually at no cost. In fact, when vendor consultation is combined with manufacturer's own engineering resources to challenge commonly accepted practices and to optimize process control over each step in an assembly line, the stage is set for creating a highly successful operation with high productivity levels (Table 1).

References

1. Zarrow, P. (1991). Fine-pitch soldering. *Circuits Assembly*. pp. 47–48.
2. Zarrow, P. (1997). Fear and loathing with double-sided reflow soldering. *Circuits Assembly*. p. 26.

Common Maintenance and Troubleshooting

Andrew Zhou

Introduction

All the furnaces have been well tested, and the parameters of the furnace have been set to right condition for user by the professional engineer. Good operation habits and well daily maintenance will extend the lifespan of furnace in a way. Here we discuss some tips of common maintenance and troubleshooting.

Common Maintenance

Daily Maintenance

- The placement of the furnace must be level and grounded firmly.
- There should be no strong convection airflow on the belt movement direction to keep the stability of furnace temperature and atmosphere inside the chamber.
- If there is some failure, the power of furnace must be cut off before checking.
- Check the condition of the internal chamber regularly; if there is any abnormality, please contact the manufacturer before taking action.
- Check the drive system regularly. Pay attention to the belt running condition. If off-tracking or slipping happened on the belt movement, please check the tension pulley and adjust the tension accordingly. After longtime operation in high temperature, the length of belt may extend a little; you may cut the redundant part once in a while.
- The process gas must be dry, clean, and oil-free.
- The heat-up operation for furnace with muffle tube will be a little bit different from normal furnace. Heating up rapidly will have a serious impact on muffle, so heating up must be slow. From the first heating up, please only heat up to 30 °C each time; when stable, please keep the temperature for over 10 min until the

J. Xu et al., *Conveyor Belt Furnace Thermal Processing*,
https://doi.org/10.1007/978-3-319-69730-7_20

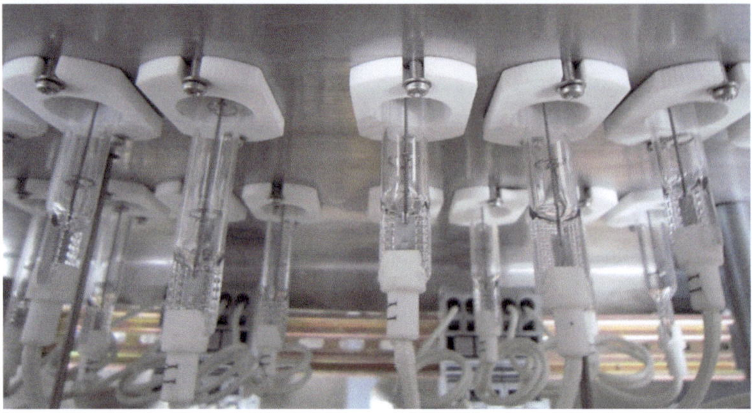

Fig. 1 Infrared lamp (quartz lamp)

temperature rises to technology profile temperature. This software can be heated up automatically; after setting the process profile, press the heating button of the furnace according to the 1 °C/min heating speed to heat up automatically.

Replacement of Heating Elements

There are three main types of heating elements in conveyor-type furnace: one is infrared lamp, one is heating board with resistance wire buried inside, and the third one is ceramic rod winded with heating wire. They are known as heating elements. If some heating elements are damaged, replace them according to the following steps:

- Turn off the power. Make sure that the main power of the furnace must be disconnected.
- Be sure that the temperature of the furnace decline to room temperature.

(a) Infrared lamp (quartz lamp) (Fig. 1)

- Remove the leading-out wire of quartz lamp tube on the ceramic wire holder (or copper bar).
- Reinstall the new lamp tube on the furnace according to the mark.
- Connect the leading-out wire to the quartz lamp tube on the WAGO terminal.
- Test the heating up by hand after power on to check the working condition of the lamp tube.

(b) Heating board (Fig. 2)

- Remove the top cover, the panels at both sides of the furnace.

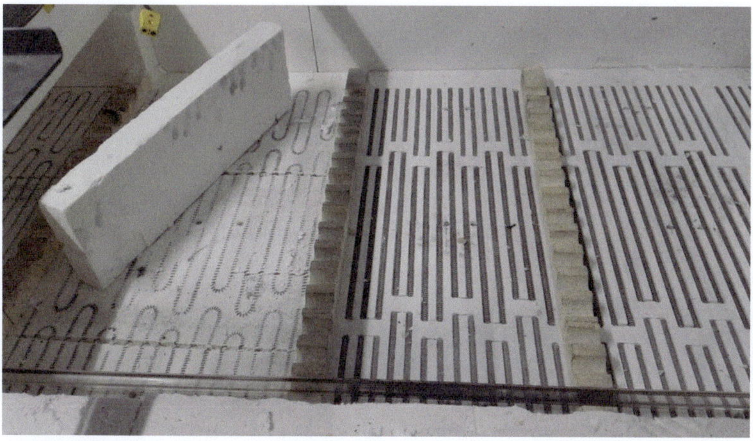

Fig. 2 Heating board

- Remove insulation protective panels outside the connection wire.
- Loose connecting wire on the damaged heating board. Remove the porcelain bowl on the connecting wire.
- Remove the corresponding thermocouple at the other end of the furnace and lateral connecting plate on the furnace which is used for insulation and fixing the thermocouple.
- Remove side insulation plates and ceramic fiber materials one by one.
- Pull out the damaged heating board from the top of the furnace.
- Replace the heating board and reinstall all side insulation plates according to reverse steps above.
- Install the lateral connecting plates and thermocouples, and connect compensating wire of the thermocouple correctly.
- Install the insulating porcelain bowl to the connecting wire of the heating elements, and then connect the connecting wire to the copper bar.
- Check the heater with avometer to ensure connection is made correctly.
- Heat up after power on and check the working condition of heaters.

(c) Ceramic rod winded with heating wire (Fig. 3)

- Remove the panels at both sides of the furnace.
- Remove insulation protective panels outside the copper bar.
- Loose connecting wire on the damaged heating elements. Remove the fiberglass pipe on the connecting wire.
- Pull out the damaged heating elements from the sides of the furnace.
- Replace the heating elements and reverse install them according to the remove steps.
- Install the fiberglass pipe on the heating element, and then connect the connecting wire of the heating wire to the copper bar.
- Check the heater with multimeter to ensure connection is made correctly.

Fig. 3 Ceramic rod winded with heating wire

- Do manual heating up after power on and check the working condition of heaters.

Maintenance of the Motor

- Keep operating environment dry, and the surface of the motor must be kept clean.
- If the thermal protection or short circuit protection of the motor continuous action occurs, please find out the failure reason (starting load of the motor is big or set value of the protective device is too low); run it again after resolving the error.
- Make sure the motor has good lubrication during running. This motor selects single-row centripetal ball bearing with sealing. The bearing is lubricated in manufacturing. Generally, after the motor has run for around 5000 h, check the radial clearance of the bearing.
- Disassembly of the motor: remove the cover of the motor first, and then pull out the rotor directly from the front. Please pay attention to prevent the stator windings and insulation and sealing washer from damage.

Transportation

Do not move the furnace at random during the use. If it needs to be transported to other places, it is necessary to do the following steps:

- Remove the thermocouple and keep it well.
- Remove the heater, mark it, and keep it well.
- Mark the joint of the furnace and then disassemble it.

Equipment Storing Up

If this furnace is not used because of some reason (non-damage), you should do the following preparation before working on storing up the furnace:

- Oil movable parts of the hinge.
- Clean the surface of the equipment.
- Wrap the entrance and exit of the chamber with airtight plastic.
- Make sure all outside panels are installed.
- Disassemble the components of the computer and keep it safely.
- Wrap the whole equipment with a plastic bag.
- Keep the equipment in a dry and cool area.

Storage of Motor

- Keep motor dry during storage and transportation to prevent rainwater to flow into the motor and to avoid the rapid change of ambient temperature.
- The motor is not suitable for stacking during storage and transportation. It will damage the package of the motor.
- The motor should not be placed upside down or inclined during storage and transportation.
- Because the motor shaft is projecting, please prevent the motor shaft against deformation caused by outside force.

Troubleshooting

Some components of the equipment start to malfunction after the long-term run. The most common heating fault phenomenon and solutions to these problems are stated below.

Overheating

Phenomenon: Overheating of a certain zone; the temperature of this zone is getting high and high. At the same time, the output of the temperature controller of this zone is 0.

Solution: Use the avometer to measure the voltage between the bottom of the solid relay of the abnormal heating zone and neutral wire to confirm whether the voltage is 220 V. If the voltage is 220 V, it means this solid relay is broken down;

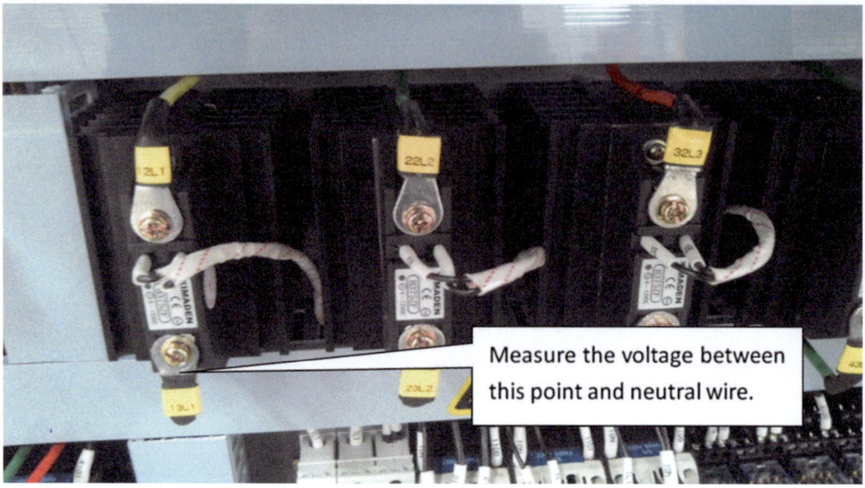

Fig. 4 The accurate measure position when overheating

replace it with a new one, and the problem will be fixed. If the voltage is not 220 V, it means this solid relay is not broken. Go and check the thermocouple and extension wire of thermocouple (Fig. 4).

Heating Up Problem

Eliminate stupid mistakes: no conduction of corresponding circuit breaker, AC contactor, and solid relay.

The measurement method is the same as stated above; use the avometer to measure the voltage between the bottom of the solid relay of the abnormal heating zone and neutral wire. If voltage measured is 0 V, and there is output power of temperature controller, the solid relay is broken.

Heating Elements Broken

If the solid relay of the abnormal heating zone is not broken, then check the corresponding heating elements whether it is good (Fig. 5).

If the current measured is not the same as the calculated current, then the heating element is broken. Replace the corresponding heating element.

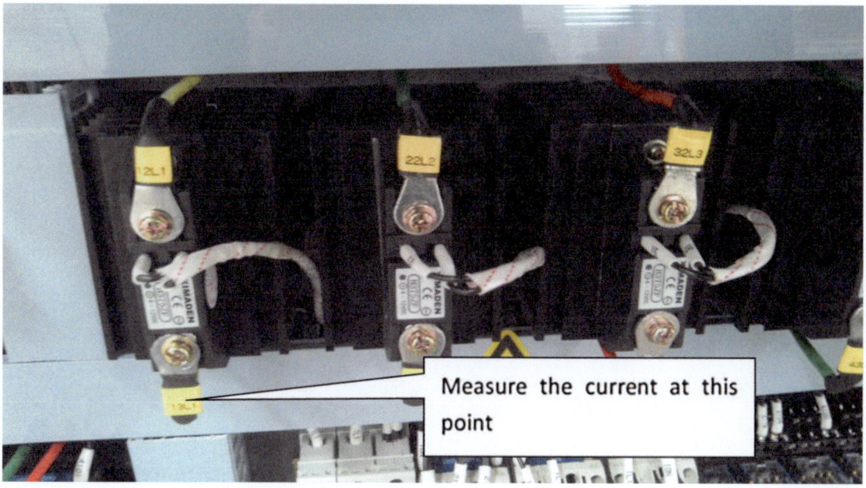

Fig. 5 The accurate measure position if the solid relay of the abnormal heating zone is not broken

Fig. 6 The tension roller

Belt Movement Problem

If belt slip happens after restoring from emergency stop, and the motor is still rotating as normal, please check the tension roller in the picture below and tighten the screw to adjust the tension (Fig. 6).

If the problem happens to the frequency converter, it will initiate VFD fault alarm on the operation panel of the furnace. Check the alarm code and find the solution in the manual of the VFD (Fig. 7).

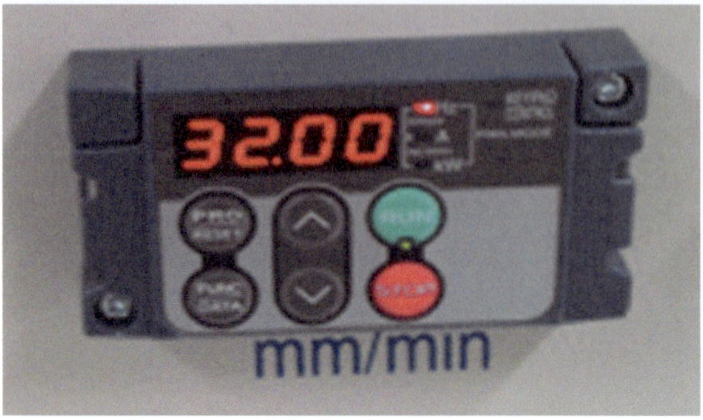

Fig. 7 Variable frequency drive (VFD)

Breaker Tripping

The capacity of breaker is not too large; if the corresponding load capacity is bigger than the breaker capacity, it will cause the tripping of the breaker. Just replace the breaker with a more fit one.

Breaker load short circuit occurs, when the load (heating element) is short-circuited with peripheral conductive objects, causing tripping of the circuit breaker; you need to find the heating elements of the corresponding zone; use the avometer to check which element is short-circuiting with peripheral conductive objects.

Black Screen of Industrial Computer Display

If the black screen of industrial computer display happens, check the VGA wire whether it is broken. If the VGA wire is okay, then check the GPU of the computer. If there are some current disturbances around the computer, use the anti-interference film to protect the VGA wire (Table 1).

Table 1 Some common problems with their causes and solutions

Problems	Possible causes	Suggested solution
The power is switched on but the power indicator lamp does not work	Main power is disconnected or the light of the power is broken	Connect the power or replace light bulb of power
The temperature controller displays wrong data or cannot display after power on	The temperature controller has bad connection or failed	Call technical personnel to see and repair
The belt does not work after pressing the belt running button	1. The yellow emergency stop switches on both ends of the furnace are pressed down	1. Reset the yellow emergency stop switch
	2. Power circuit of the motor has failed	2. Call technical personnel to repair

Problems	Possible causes	Suggested solution
When the furnace begins to heat, the temperature of one zone or two zones increases so slowly or cannot rise	1. One or two switches are disconnected	1. Turn on the main power switch
	2. Corresponding solid state relay has failed	2. Replace solid state relay
The temperature of some zone rises to over 100 °C and still continuously rising when it starts to heat up	This solid state relay has a short circuit	Replace solid state relay

Future Development

Yanjun Cheng

Introduction

Current development of industrial manufacturing is indeed remarkable, while more and more improvements have been promoted and applied. New advancements, including new materials and technology innovations, have great influence on the heating processes. Moreover, the optimized steps in heating processes also enhance the quality of products and efficiency of fabrications. Further exploitation will be mainly developed in these trails and may spread in other branches as well.

New Materials Result in Different Thermal Profiling

In recent years, many kinds of new materials have been introduced into different industrial processes. Those new materials with distinctive properties have various influences to the fabricating methods. Some examples will be mentioned in the following passage. Thus, in this pattern, it is predictable that more and more new materials will be tested and used in industrial fields in the future and the thermal processes will be more effective and maybe simpler according to the materials used.

Solar Cells

Besides silicon, thin film, and dye-sensitized solar cells, perovskite solar cells have attracted increasing attention these years for their high efficiencies over 20%. The perovskite structured compound served as the active absorber of light. The most common material for perovskite solar cells is a hybrid organic-inorganic lead or tin halide-based material which is low cost and high productive (Fig. 1).

J. Xu et al., *Conveyor Belt Furnace Thermal Processing*,
https://doi.org/10.1007/978-3-319-69730-7_21

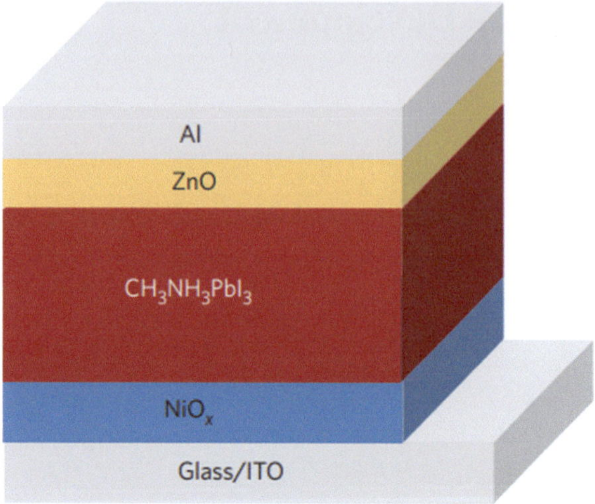

Fig. 1 Structure of perovskite solar cells

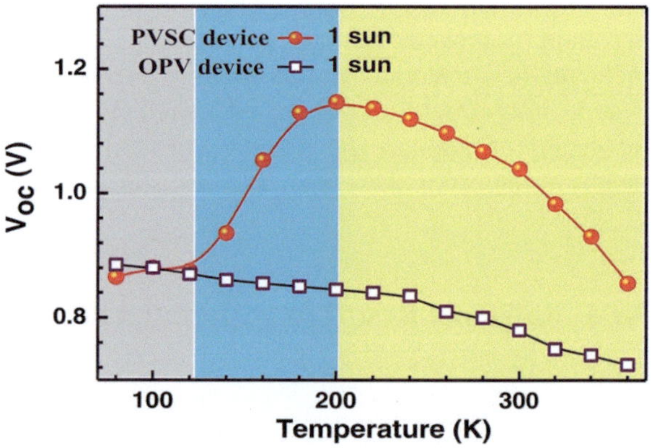

Fig. 2 Efficiency of the $CH_3NH_3PbI_3$-based perovskite solar cells under temperature from 80 to 360 K

Also, the manufacturing processing is much simpler than that of the traditional silicon solar cells. Traditional silicon solar cells need to be conducted at temperatures above 1000 °C, while perovskite solar cells can be prepared at a lower temperature due to the organic component and strong ionic interactions among those materials. For example, a study of the $CH_3NH_3PbI_3$-based perovskite solar cells reveals that because of the ferroelectric-paraelectric phase transition performed by $CH_3NH_3PbI_3$ when temperature is lower than 160 K, this kind of solar cells exhibits strong temperature-dependent photovoltaic effects [1]. When temperature is lower than 120 K, the charge transport layer reduces the efficiency of the device, while when the temperature is higher than 200 K, the interfacial charge recombination acts as the decisive factor of cell's efficiency (Fig. 2) [1].

Engine Valves

As mentioned in the previous chapter, engine valves are critical parts for engine functioning. Engine valves can be divided into intake valves working at temperature 427–538 °C and exhaust valves working at temperature 649–788 °C, due to the different temperatures of intake and exhaust gas. To fulfill the tough working condition of high temperature and provide enough strength and hardness, heat-resistant steels are the proper choices for engine valves. At the same time, various alloys are found to improve the efficiency of engine valves. For the exhaust valves, several alloys have been introduced and tested by G. Muralidharan [2]. The temperature tolerance of the new material is largely improved so that they can operate at 870 °C [2]. New alloys are proved to have higher strength and ductility compared to other alloys at 870 °C. Moreover, traditional alloys for exhaust valves are high cost because of the high proportion of Ni and Co. These new alloys reduce the ratio of Ni and Co [2]. Instead, they use commercial alloys with desirable microstructure to maintain and advance the functionality.

Brazing Process

Technological progress has expended the scope of brazing process from joining metals together to connecting other materials such as glass and ceramics. Brazing is an essential process in many industrial fields, such as electronics, aerospace, and transportation, to fabricate large structure and complex-shaped components. For example, brazing of SiC ceramic becomes prevalent due to the stable properties of SiC ceramic at high temperature with filler metal Ag-Cu-Ti. However, the difference of CTE (coefficient of thermal expansion) between SiC ceramic and Au-Cu-Ti contributes to the residual stresses in the interfacial region, which reduce the quality of the brazed joints. Recently, novel B_4C reinforced Ag-Cu-Ti composite filler had been introduced to braze SiC ceramics and solve this problem [3]. The addition of B_4C particles promotes the simultaneous composition of TiB whisker and TiC particles, which further reinforce the microstructure of SiC/SiC joints. Brazing temperature has great influence on the microstructure and bending strength of the joints. The desirable microstructure and maximum bending strength can be obtained by brazing at 950 °C for 10 min, and the strength will drop dramatically if the temperature keeps going up (Fig. 3) [3].

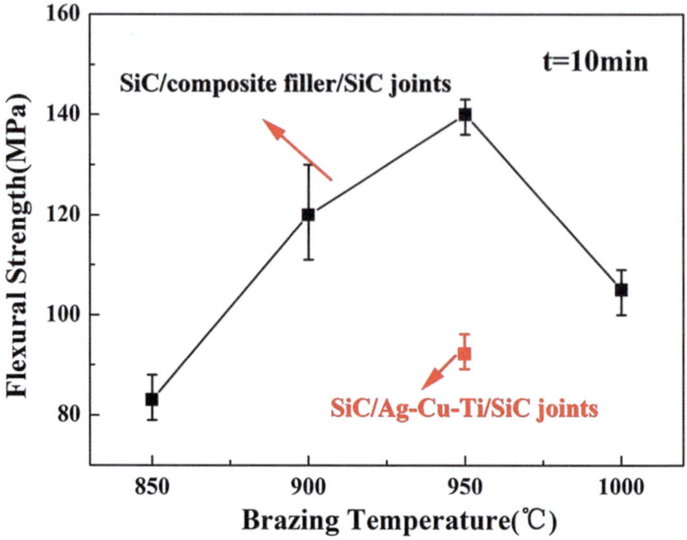

Fig. 3 Bending strength of the joints under different temperatures for 10 min

Technology Innovations and Their Influence on Thermal Processes

Not only the new materials trigger the adjustments of the thermal processes, but the innovations in different fabricate steps also influence the industrial heating process in some ways. Recent improvements include the optimization in thick film, glass-to-metal seal process, DBC, brazing process, and so on. Further innovations will come forth definitely, and because of these innovations, the thermal process parameters will be more accurate and the product strength will also get improved.

Thick Film

Thick film is a technology that fuses pastes or inks onto a ceramic substrate permanently at a high temperature of 500–1000 °C. The firing process includes organic burnout, softening or melt, resistive compound formation, solidification, and annealing successively. With different formulas, the pastes can be used to produce conductors, resistors, and dielectrics with substrates. An experiment investigated the effect of annealing process on the thermoelectric properties of the screen-printed Bi_2Te_3 thick film [4]. It illustrates that the annealing time, temperature, and environment are all critical factors in determining the quality of the thick film. The results of this experiment show that the printed Bi_2Te_3 thick film reached a power factor of 2.1 mW/mK^2, a thermal conductivity of 1.0 W/mK, and a higher ZT value when annealed at 500 °C for 15 min in Bi and Te powder ambient [4]. Thus, the optimization of the annealing process largely increases the thermoelectric properties of the screen-printed Bi_2Te_3 thick film and makes it a proper choice for low-cost and large-size thermoelectric device operating at room temperature.

Glass-to-Metal Seal Process

Glass-to-metal seal process is a common technology in the industrial fabrication for remarkable properties and reliability. Inorganic glass is heated, wetting the metal in order to form a strong bond when the temperature drops back. The rates and coefficients of the glass and the metal should be close or similar. By combining the anodic bonding and a liquid solder, a firm connection between the bonding surfaces of the glass/solder/metal seals is created. While the anodic bonding across the oxide/metal surface enhances the glass/solder bond, the soft soldering increases the strength of the solder/metal bond. With the addition of Al to the solder, the bond strength between glass and solder is improved with increasing proportion of Al content. However, at the same time, the strength of solder/metal bond decreases. Recently, researchers have found that a low Al content of 75 ppm offers an intermediate [5]. Moreover, the optimal composition of solder improves the strength of glass/solder/metal bonds much more than the solder with no Al addition.

DBC Technology

DBC (direct bonded copper) technology consists of a ceramic substrate (usually Al_2O_3) with bonded copper on one or both sides. It offsets the disadvantages of thin and thick film by the stability in high-current, high-voltage, high-power, and high-temperature environment. To extend the lifetime of the DBC substrates, a new structure with ladder-shaped copper layers has been proposed. Experiments verified that the lifetime of ladder-shaped Al_2O_3 substrates is almost three times of the traditional ones (Fig. 4).

In order to achieve a longer lifetime, the substrate should reduce the influence of plastic strain distortion as much as possible. For the ladder-shaped substrate, when the thickness of the layers both equals to 0.15 mm, the equivalent plastic distortion reaches its minimum [6]. The length of the edge tail is also a determining factor of plastic strain distortion, especially the length of the edge tail of the second copper layer, while that of the first copper layer influenced the warpage [6]. Conclusion also includes that prolonging the tail length increases the lifetime of the ladder-shaped DBC substrates [6].

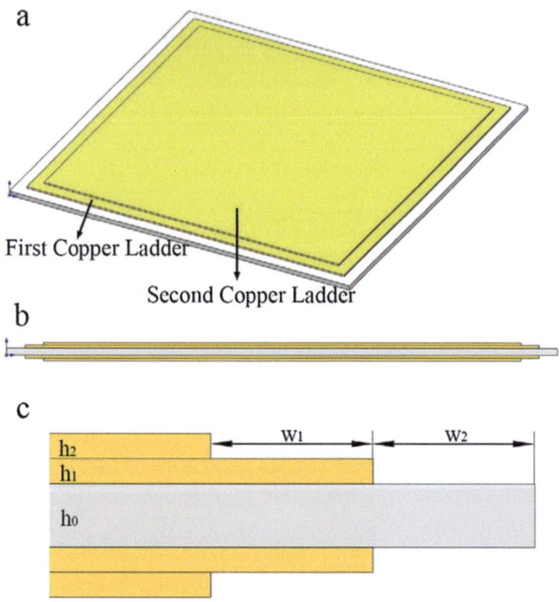

Fig. 4 Ladder-shaped DBC substrate. (**a**) top view; (**b**) side view; (**c**) close vieq

Brazing Process

Vacuum brazing technology is prominent for its strong joints and less distortion. The temperature of the base metal is much stable. The residual stresses also get reduced because of the slow heating and fast cooling cycle. Thus, mechanical and thermal properties of vacuum brazing products are remarkably improved. Therefore, it is significant to determine the brazing parameters during vacuum brazing process, and Taguchi method is a good choice to optimize the parameters. By using Taguchi method, the optimal brazing parameters for soaking time, soaking temperature, brazing temperature, and brazing time to reach highest tensile strength are 590 °C, 50 min, 600 °C, and 30 min [7]. Under this circumstance, the error between the expected value and experimental value of tensile strength is only 0.33% [7]. As a result, Taguchi method is a feasible and optimal way to find vacuum brazing parameters in industrial fabrications.

Optimizing the Heating Process

Heating process is one of the most common and crucial steps in many branches of the industrial manufacture. Optimizing the heating process is a main idea of the future development of fabrication. To control the heating temperature more precisely, to boost the productivity, to determine optimal line speed of the furnace, and to clean the furnace thoroughly are all popular topics and advance in heating process.

Precise Temperature Control

In an automatic control system (ASC), controller is a critical part to keep the system working precisely, and PID is one of the most frequently used algorithms for a controller. To find the best parameters for the controller, controller tuning is the necessary and vital process. There are many ways of controller tuning including theoretical, numerical, and experimental approaches. For example, a recent research presents a method of optimizing tuning parameters based on minimizing integral performance indices by using MATLAB software [8]. It performs two case studies considering a proportional controller and a proportional-integral controller, respectively, and finds the optimal values for both cases [8]. Besides the one presented before, the other algorithm like probabilistic or even artificial intelligence has also been applied to find optimal value for controller and thus promotes more precise temperature controlling in the heating process.

Mass Production

To fulfill the mass production mandatory in industry manufacturing, it is necessary to optimize the efficiency of heating process. With oxygen-enhanced combustion technologies applied to reheating furnace, gas missions are reduced, while thermal efficiency and productivity are increased [9]. The control of the oxygen combustion technologies can be optimized by mathematical model equations, thus to develop an eco-friendly and high efficient reheating furnace (Fig. 5).

Improving the efficiency of cool down process can also promote the productivity. The time for cooling down can be reduced by 30% by introducing mass flow through the furnace [10]. Optimal mass flow is characterized by a nonlinear, radially symmetric model of furnace [10]. Adopting this optimal mass flow to furnace not only cuts down the cooling time and rises the efficiency but also maximizes the profit from the sintering process.

Line Speed

Traditional furnace has to set up the line speed by experienced operators. For different operators or even on different machines, the parameters may result in totally different products. It is really difficult to keep the setting and time of changing of the line speed uniform on every furnace. To solve this problem, JFE Steel has introduced the plant optimal control system and applied it on the No. 5 Continuous Annealing Line (5CAL) [11]. With this system, the line speed is able to set up automatically based on the precise calculation. It eliminates the errors and deviations in the line speed setting by operators based on their experience. Moreover, a strip

Fig. 5 Reheating furnace of oxygen-enhanced combustion

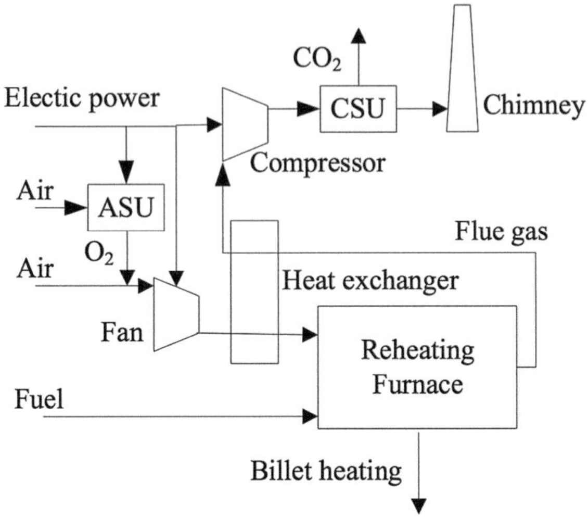

temperature feedforward function is also developed to minimize the gaps of strip temperature between the upper and lower limits [11].

Cleaning

Besides the ultrasonic belt cleaning from the previous chapter, many cleaning methods are applied on different furnaces. During the analytical process, residue and dust will stick to the combustion tube and filter. Present designs have a vacuum dust door for collecting the contaminations, but sometimes, the combustion crucible falls down and blocks the vacuum dust door. An invention of combustion furnace auto cleaner is newly published to solve this problem. This invention providing an automatic cleaning assembly is attached on the filter chamber [12]. When a vacuum is drawn through the lower end of the combustion tube, a brush of the cleaning assembly rotates through the filter chamber and the combustion tube [12]. This enlarges the pressure difference and accelerates the cleaning rate. The auto cleaning system improves the productivity and quality of the analytics by cutting down the time for dealing with the crucible blocks and manual cleaning.

References

1. Zhang, H., Qiao, X., Shen, Y., & Wang, M. (2015). Effect of temperature on the efficiency of organometallic perovskite solar cells. *Journal of Energy Chemistry, 24*, 729–735.
2. Muralidharan, G. (2013). *Exhaust valve materials for high efficiency engines*. https://energy.gov/sites/prod/files/2014/03/f13/pm047_muralidharan_2013_o.pdf

3. http://www.sciencedirect.com/science/article/pii/S0272884216000493
4. We, J. H., Kim, S. J., Kim, G. S., & Cho, B. J. (2013). Improvement of thermoelectric properties of screen-printed Bi_2Te_3 thick film by optimization of the annealing process. *Journal of Alloys and Compounds, 552*, 107–110.
5. Malfait, W. J., Klemencic, R., Lang, B., Rist, T., Klucka, M., Zajaczc, Z., & Koebel, M. M. (2016). Optimized solder alloy for glass-to-metal joints by simultaneous soldering and anodic bonding. *Journal of Materials Processing Technology, 236*, 176–182.
6. Xu, L., Wang, M., Zhou, Y., Qian, Z., & Liu, S. (2016). An optimal structure design to improve the reliability of Al_2O_3-DBC substrates under thermal cycling. *Microelectronics Reliability, 56*, 101–108.
7. Lin, H., Hwang, J. R., & Fung, C. P. (2016). Optimization of vacuum brazing process parameters in AA6061 using Taguchi method. *Journal of Advanced Mechanical Design, System, and Manufacturing*, 1–10.
8. Doicin, B., Popescu, M., & Patrascioiu, C. (2016). *PID controller optimal tuning*. ECAI International Conference (8th edn, pp. 49–52). Ploiesti.
9. Wang, N., Wen, Z., & Su, F. (2013). *Optimal heating and energy management for the reheating furnace using oxygen enhanced combustion*. AICHE Proceeding.
10. Rogers, A. P., & Rasmussen, B. P. (2017). *Determining the optimal cool-down for an industrial sintering process*. American Control Conference. (pp. 2207–2212). Seattle.
11. Takurou, M., Tatsuo, T., & Masayasu, O. (2016). *Plant optimal control system for No. 5 continuous annealing line (CAL) at West Japan Works (Fukuyama), JFE Steel. JFE Technical Report* (pp. 112–115).
12. Ford, G. C., Wetzel, J. N., & Brushwyler, K. R. (2011). Combustion furnace auto cleaner, US Patent Application, US9541287.

Index

J. Xu et al., *Conveyor Belt Furnace Thermal Processing*,
https://doi.org/10.1007/978-3-319-69730-7

MIX
Papier aus verantwortungsvollen Quellen
Paper from responsible sources
FSC® C105338
FSC
www.fsc.org

If you have any concerns about our products,
you can contact us on
ProductSafety@springernature.com

In case Publisher is established outside the EU,
the EU authorized representative is:
Springer Nature Customer Service Center GmbH
Europaplatz 3, 69115 Heidelberg, Germany

Printed by Libri Plureos GmbH
in Hamburg, Germany